Corn and Grain Sorghum Comparison

Corn and Grain Sorghum Comparison
All Things Considered

Yared Assefa
Department of Agronomy, Kansas State University, Manhattan, KS

Kraig Roozeboom
Department of Agronomy, Kansas State University, Manhattan, KS

Curtis Thompson
Department of Agronomy, Kansas State University, Manhattan, KS

Alan Schlegel
SW Research-Extension Center, Kansas State University, Tribune, KS

Loyd Stone
Department of Agronomy, Kansas State University, Manhattan, KS

Jane E. Lingenfelser
Department of Agronomy, Kansas State University, Manhattan, KS

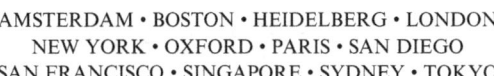

AMSTERDAM • BOSTON • HEIDELBERG • LONDON
NEW YORK • OXFORD • PARIS • SAN DIEGO
SAN FRANCISCO • SINGAPORE • SYDNEY • TOKYO

ELSEVIER   Academic Press is an imprint of Elsevier

Academic Press is an imprint of Elsevier
The Boulevard, Langford Lane, Kidlington, Oxford, OX5 1GB, UK
225 Wyman Street, Waltham, MA 02451, USA

First published 2014

Copyright © 2014 Elsevier Inc. All rights reserved.

No part of this publication may be reproduced or transmitted in any form or by any means, electronic or mechanical, including photocopying, recording, or any information storage and retrieval system, without permission in writing from the publisher. Details on how to seek permission, further information about the Publisher's permissions policies and our arrangement with organizations such as the Copyright Clearance Center and the Copyright Licensing Agency, can be found at our website: www.elsevier.com/permissions

This book and the individual contributions contained in it are protected under copyright by the Publisher (other than as may be noted herein).

Notices
Knowledge and best practice in this field are constantly changing. As new research and experience broaden our understanding, changes in research methods, professional practices, or medical treatment may become necessary.

Practitioners and researchers must always rely on their own experience and knowledge in evaluating and using any information, methods, compounds, or experiments described herein. In using such information or methods they should be mindful of their own safety and the safety of others, including parties for whom they have a professional responsibility.

To the fullest extent of the law, neither the Publisher nor the authors, contributors, or editors, assume any liability for any injury and/or damage to persons or property as a matter of products liability, negligence or otherwise, or from any use or operation of any methods, products, instructions, or ideas contained in the material herein.

British Library Cataloguing in Publication Data
A catalogue record for this book is available from the British Library

Library of Congress Cataloging-in-Publication Data
A catalog record for this book is available from the Library of Congress

ISBN: 978-0-12-800112-7

For information on all Academic Press publications
visit our website at http://store.elsevier.com

This book has been manufactured using Print On Demand technology. Each copy is produced to order and is limited to black ink. The online version of this book will show color figures where appropriate.

Transferred to Digital Printing in 2013

CONTENTS

Acknowledgments vii

Chapter 1 Introduction 1

Chapter 2 Corn and Grain Sorghum Morphology, Physiology, and Phenology 3
2.1 Morphology 3
2.2 Physiology 10
2.3 Phenology 12

Chapter 3 Corn and Grain Sorghum Historical Yield, Area, and Price Trends 15
3.1 Analytical Procedures 16
3.2 Harvested Area 17
3.3 Grain Yield 21
3.4 Prices of Corn and Sorghum, 1949–2011 28

Chapter 4 Corn and Grain Sorghum Yield Trend Since the Beginning of Hybrid Technology 31
4.1 Introduction 31
4.2 Corn from 1939 to 2009 33
4.3 Grain Sorghum from 1957 to 2009 48
4.4 Conclusion 55

Chapter 5 Yield Distribution and Functions for Corn and Grain Sorghum 57
5.1 Data Assembly and Analytical Steps 57
5.2 General Yield Distribution 59
5.3 Partitioning Sources of Variability 60
5.4 Environment Comparison 62
5.5 Corn and Grain Sorghum Yield Models 63

**Chapter 6 Resource (Land, Water, Nutrient, and Pesticide)
Use and Efficiency of Corn and Sorghum 71**
6.1   Land Use Efficiency (Yield) 72
6.2   Water Requirements and Water Use Efficiency 78
6.3   Fertilizer Requirements and Fertilizer Use Efficiency 81
6.4   Pesticide Requirements and Use Efficiency 83
6.5   Preliminary Economic Considerations (General Resource
        Use Efficiency) 85

**Chapter 7 Rotation Effects of Corn and Sorghum in Cropping
Systems 87**
7.1   Analytical Approach 87
7.2   Survey of Common Rotations and Factors Affecting Crop
        Sequencing Decisions 90
7.3   Analysis of Rotation Studies 93
7.4   Analysis of Crop Sequencing Relative to Crop Water Use
        and Optimum Planting and Harvest Dates 97

Chapter 8 General Summary 103

References 107

ACKNOWLEDGMENTS

This research was supported by the Kansas Corn Commission and partly by the Kansas Sorghum Commission. We would like to thank Dr. Daniel O'Brein, Dr. Mykel Taylor, and Marcus Brix of Kansas State University, Department of Economics, for comments at different steps when this research was developed, and Dr. Jason Lamprecht (Senior Statistician, USDA-NASS, KS Ag Statistics) and Dr. Randall Nelson (Assistant Professor, North Central Experimental Field Manager) for sharing information and data. This is contribution 14-002-B from the Kansas Agricultural Experiment Station.

CHAPTER 1

Introduction

Corn and grain sorghum are among the top cereal crops worldwide, and both are key for global food security. Both grain sorghum and corn belong to the grass family *Poaceae* (Gramineae). Members of this family of plants are considered to be the most important of all plant families due to their economical and ecological significance (Campbell, 2012; Zhang, 2000). Within the family of *Poaceae*, corn is classified in the genus *Zea*, along with grasses native to Mexico and Central America (CFIA, 1994), and sorghum is classified in the genus *Sorghum* and is native to Ethiopia in the Horn of Africa (Dillon et al., 2007a; Wet and Harlan, 1971).

Corn grows almost everywhere in the world, with most extensive production in regions approximately between 50°N and 45°S, whereas sorghum production is restricted to arid and semiarid regions of the world (Leff et al., 2004; Dillon et al., 2007b). The principal arid regions of the world and crops adapted to these regions, including corn and sorghum, are described in the paper by Creswell and Martin (1998). Creswell and Martin defined arid regions as those with long dry seasons or where potential evapotranspiration of water exceeds rainfall. The three categories of arid regions are: (i) regions with dry climate but with seasonal rain, (ii) regions with all-around aridity modified by light or irregular rain, and (iii) regions with all-around aridity where water is brought only with wells, canals, or other means. Much of the western United States constitutes one of the larger arid regions of the world.

The similarities between corn and grain sorghum, particularly the fact that they are warm-season cereals with a C_4 pathway adapted for summer season cropping in the United States, have made them a topic for comparison by many authors (Gordon and Staggenborg, 2003; Norwood, 1999; Martin, 1930; Mason et al., 2008; Staggenborg et al., 2008; Stone et al., 1996). In places where rainfall is dependable or irrigation systems are available, corn is the dominant summer crop due to yield superiority. In places where rainfall amount and timing is not dependable—in parts of Oklahoma, Kansas, Colorado, Texas, and Nebraska, for example—sorghum has been historically recommended. Due to reasons such as changes in the price relationship of the two crops, release of new hybrids, availability of pesticides, and the like, however, the area allocated to sorghum in the drylands of the United States has been declining since the early 1970s.

The main objective of this book is to provide a comprehensive review of the comparison of corn and grain sorghum. Specific objectives were: (i) to investigate and document key morphological, physiological, and developmental characteristics of corn and grain sorghum in comparison to one another, (ii) to investigate, compare, and document historical trends for grain sorghum and corn yield, harvested area, and price, (iii) to create better understanding of corn and grain sorghum yield distribution and major factors responsible for yield variability, (iv) to compare land, water, nutrient, and pesticide use and use efficiencies of the two crops, and (v) to evaluate the effects of corn and sorghum in crop rotations common to the Great Plains.

CHAPTER 2

Corn and Grain Sorghum Morphology, Physiology, and Phenology

Plant morphology, the study of plant anatomy or form, is a reflection of ecological adaptation (Kaplan, 2001). Due to the strong relationship between form and function, plant morphology has strong implications for plant physiology. Plant phenology, or timing of developmental events in plants (Koch et al., 2007), is also a reflection of ecological and genetic characteristics. A study of corn and sorghum morphology, physiology, and phenology is important for understanding the drivers of differential performance of these crops; however, a recent comprehensive study does not exist on this topic. The objective of this review paper was to investigate the morphological, physiological, and developmental characteristics of corn and grain sorghum to help understand reasons behind the differences and similarities between the two crops. Reports of more than 60 peer-reviewed journal articles, extension publications, books, and electronic resources were reviewed.

2.1 MORPHOLOGY

In the vegetative stage, grain sorghum and corn have a similar appearance, which can pose an identification challenge even for an expert (Figure 2.1). Root, stem, and leaves comprise the vegetative stage morphology of corn and sorghum. Despite the similar appearance of these vegetative parts of the two crops, the detailed differences in the number, size, distribution, and response of underground and aboveground vegetative parts of the two crops are among the crucial factors cited as responsible for variation in yield and adaptation.

2.1.1 Underground Morphology—Root Type, Growth, and Distribution

Embryonic roots of cereals comprise seminal roots and primary roots (Abbe and Stein, 1954; Hochholdinger et al., 2005). Seminal roots are formed at the scutellar node, a node located in the seed embryo (Hochholdinger et al., 2004). Sorghum appears to develop no seminal

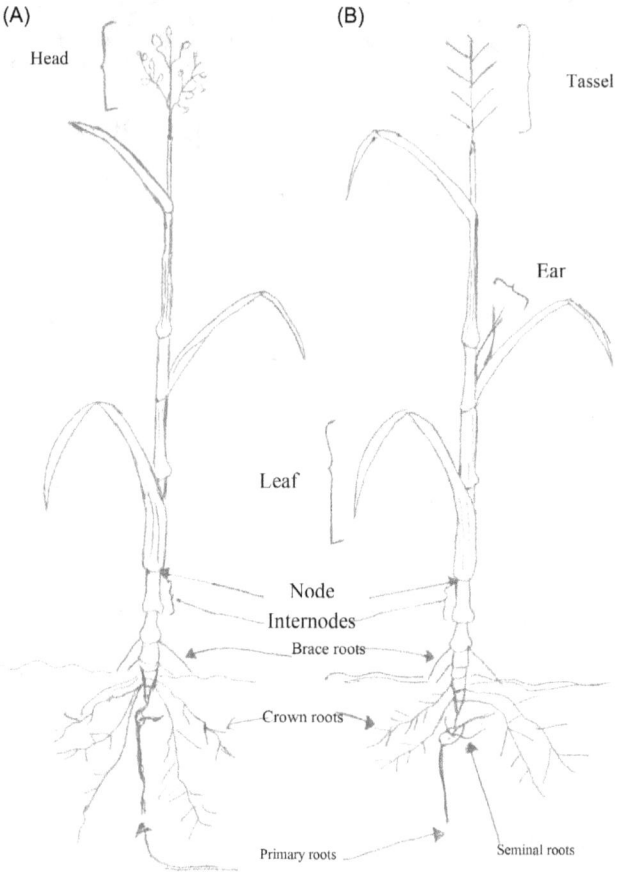

Figure 2.1 A sketch of mature grain sorghum (A) and corn (B) plants. Except for the position and form of reproductive structures, corn and sorghum are similar in the appearance of their vegetative structures.

root in its life cycle (Blum et al., 1977; Singh et al., 2010; Sieglinger, 1920). Corn usually develops three or more seminal roots with short longevity (Ritchie et al., 1996; Weaver, 1926; Singh et al., 2010). The second embryonic root called the primary root is formed from the radicle, the basal pole of the seed embryo. Based on all the references above, both corn and sorghum appear to develop one primary root. Like seminal roots, the function of the primary roots in both corn and sorghum is temporary, i.e., as the plant develops post-embryonic roots, the growth and contribution of the primary root become negligible. Primary roots of sorghum are more vertically oriented than that of corn (Figure 2.2, Singh et al., 2010). Some authors refer to seminal

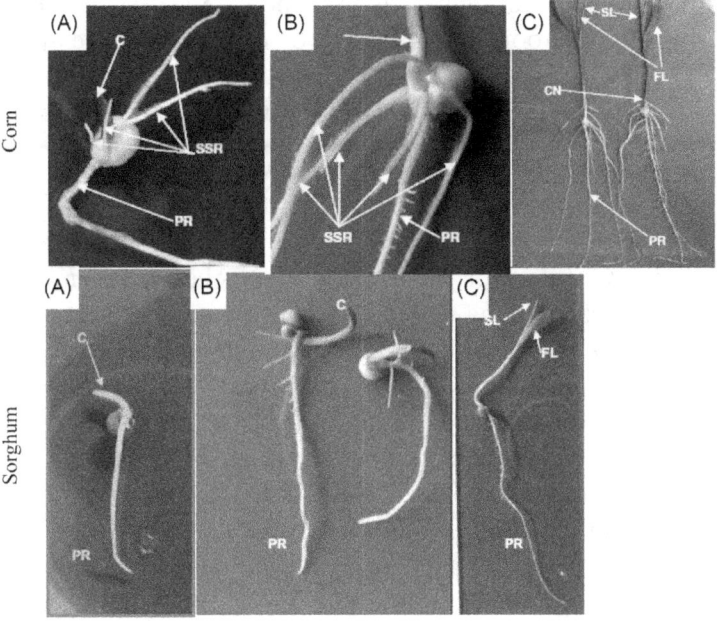

Figure 2.2 Photographs adapted from Singh et al. (2010) depicting corn and sorghum: 3 days (A), 5 days (B), and 7 days (C) after germination. C, coleoptile; CN, coleoptile node; FL, first leaf; SL, second leaf; SSR, seminal root; PR, primary root.

roots as "scutellar seminal roots" and reserve the term *seminal root* to refer the two embryonic roots (Singh et al., 2010; Watt et al., 2007).

The post-embryonic roots, or shoot-borne roots, in both sorghum and corn are called crown and brace roots. Different authors have used various terms for shoot-borne roots, such as secondary, nodal, adventitious, or permanent roots (Abendroth et al., 2011; Wenzel et al., 1989; Nishimoto and Warren, 1971; Artschwager, 1948). Crown roots are formed at stem nodes located below or at the soil surface. The first four compressed nodes and the next two elongated nodes (fifth and sixth nodes) usually produce crown roots. Brace roots are formed at nodes above the soil surface, usually up to a maximum of three nodes above the sixth node.

Early stage morphological and architectural development of the embryonic and post-embryonic roots systems of sorghum and corn was studied by Singh et al. (2010), who found that post-embryonic roots were apparent for corn as early as the second leaf stage, whereas nodal root appearance for sorghum began at the 4–5-leaf stage. Due to the absence of seminal roots (part of embryonic roots) and late appearance of post-embryonic roots in sorghum, corn appears to have more root biomass than sorghum at early growth, or up to the 6-leaf stage.

An extensive study and comparison of the root systems of corn and sorghum was conducted by Miller (1916) using two sorghum varieties and a single-corn variety in field conditions. In this study, the term *primary root* was used for the root that branches out from the node, but in the present naming system we refer to crown and brace roots. The term *secondary root* was used for the lateral roots that branched out from these primary roots. The study showed that corn and sorghum crops have the same number of primary roots at maturity (Miller, 1916); however, the primary roots of sorghum were finer and more fibrous than those of corn (Miller, 1916; Weaver, 1926). Miller (1916) also reported that sorghum has twice as many secondary roots as corn per unit of primary roots at all stages of development. Despite its age and outdated terminology, this study has been the core work referenced for sorghum and corn root system comparisons by a number of researchers. Corn also has been reported to develop up to 70 shoot-borne roots (crown and brace roots) in its life cycle (Hoppe et al., 1986).

Root distribution and activity can vary greatly depending on environmental conditions, mainly soil type, moisture, tillage, and fertilizer application (Fisher and Dunham, 1984; Blum and Ritchie, 1984). The distribution and root activity of corn and sorghum have been addressed independently and comparatively by researchers. Nakayama and Bavel (1963) reported that 90% of root activity of irrigated sorghum is concentrated within the 91-cm-deep and 38-cm-wide area from the plant. Anderson (1987) reported that corn roots sampled at the 60-cm-deep and 38-cm-wide area from the plant accounted for 82% of the total root weight collected after harvest. In a comparative study, approximately 64% and 86% of the total underground root dry matter was found in the upper 30 cm for irrigated corn and sorghum, respectively (Mayaki et al., 1976). The percentages of root dry matter at the same depth for nonirrigated corn and sorghum were 39% and 79%, respectively. From this result, we can infer that a greater percentage of corn roots were found below 30 cm compared with sorghum; however, corn roots were not found deeper than 1.5 m, whereas research showed that sorghum roots can extend up to 2.5–3 m (Shackel and Hall, 1984; Stone et al., 2002; Moroke et al., 2005). These results suggest not only that a greater percentage of sorghum roots are in the upper soil profile compared with corn, but also the remaining roots penetrate much deeper than corn.

In conclusion, corn and sorghum appear to vary in number, type, orientation, total biomass, distribution, and size of their root system. Unlike sorghum, corn has multiple seminal roots as part of its embryonic root system, and post-embryonic roots appear and develop faster and earlier in its growth. Unlike corn, sorghum, on the other hand, has simple embryonic root system that constitutes only a vertically oriented primary root and lateral branches. In addition, sorghum has the highest number of post-embryonic roots at later growth stages, with the highest percentage on surface, and these roots penetrate deeper than corn's.

2.1.2 Aboveground Morphology—Plant Height, Stem, and Leaf

The plant heights of typical corn and sorghum hybrids in different Kansas agricultural districts are presented in Table 2.1. In both irrigated and dryland cropping systems, corn has taller stalks than

Table 2.1 Average Heights of Corn and Sorghum in Dryland and Irrigated Cropping Systems in Different Districts of Kansas

District	Plant Height of Corn (in.)		Plant Height of Sorghum (in.)	
	Dryland	Irrigated	Dryland	Irrigated
Northeast	105	–	50	–
East central	111	113	49	–
Southeast	106	–	54	–
North central	90	114	44	53
Central	72	108	43	–
South central	87	103	47	57
Northwest	–	104	39	50
West central	83	103	43	50
Southwest	83	103	42	50
Average	92	107	46	52

Source: *Average from data reported on Kansas Corn and Grain sorghum performance trial data for the years 2004 through 2008.*

sorghum. Irrigated cropping systems of both crops tend to have taller stalks than dryland counterparts.

Stems consist of nodes and internodes. In both corn and sorghum, as is the case for all grasses, the number of leaves and the number of nodes are identical. A typical corn plant in the central US has 20–24 nodes and internodes (Kiesselbach, 1999; Frank, 2010). A typical sorghum plant in Kansas has 14–17 nodes and internodes. In both crops, the first four to six internodes are compressed and remain below or near the soil surface. Based on plant height and numbers of node or internodes, we can conclude that a typical corn plant has more leaves than a typical sorghum plant in the United States. A relatively large plant height and larger number of leaves translate into higher interception of photosynthetically active solar radiation and then to yield (Scarsbrook and Doss, 1973).

Leaf margins are smooth (entire) for corn and are toothed (finely serrated) for grain sorghum, a feature that has been used in identification during vegetative stages (Hannaway and Myers, 2004; Martin, 1920). Several studies have shown that the number of leaves for both sorghum and corn plants varies by environment and genotype (Hesketh, et al., 1969; Quinby et al., 1973; Sieglinger, 1936; Warrington and Kanemasu, 1983). A study of the world collection of genotypes of corn showed that the number of leaves for corn varied from 8 to 48 leaves per plant depending on the variety and the environment where they were adapted (Kuleshov, 1933). The average number of leaves in the corn varieties well adapted to the United States at that time ranged from 12 to 25. A review of the number of leaves of corn as affected by temperature and photoperiod reported by Warrington and Kanemasu (1983) showed a similar range in number of leaves (12–28). The total average number of leaves counted per plant from 21 varieties of sorghum by Sieglinger (1936) was 16–27. A sorghum plant well adapted to temperate regions may have 14–17 leaves, but as many as 30 leaves can be possible in photoperiod-sensitive hybrids (House, 1985).

Stomata number and size on the leaves of sorghum and corn also differ. Martin (1930) reported that the size of sorghum stomata was two-thirds the size of corn stomata, but the number of stomata in a sorghum leaf was twice that of corn. In corn, the number of stomata per square millimeter of a leaf area varies from 60 to 80 (Kramer and

Boyer, 1995). The upper leaf surface of both crops contains two-thirds as many stomata as the lower surface has. Studies conducted to examine the relationship between stomatal size and frequency and yield or transpiration within species of crops have had mixed conclusions; some suggest positive relationships, whereas others find negative or no relationships (Muchow and Sinclair, 1989; Teare et al., 1971; Maghsoudi and Moud, 2008). However, different responses of stomata for different water-deficit conditions were documented for corn and sorghum. Sorghum has the ability to maintain stomatal openings at low levels of water potential and under a wide range of leaf turgors (Assefa et al., 2010). Sorghum stomata can remain open over a wider range of leaf turgor than corn (Turner, 1974; Sanchez-Diaz and Karmer, 1971). This adaptation enables sorghum to maintain a higher rate of CO_2 exchange than corn at a high level of water stress.

2.1.3 Reproductive Morphology of Corn and Sorghum

The reproductive structures of corn and sorghum are a source of clear and visible morphological difference between the two crops. Corn is a monoecious crop with incomplete and imperfect flowers, i.e., its stamen or male floral part and the pistil or female floral part are positioned at different locations in the plant (Figure 2.1; Weatherwax, 1916). The stamens are located on the tassel, which is the inflorescence located on a peduncle above the flag leaf. The pistil is located on the ear, which is an inflorescence located at the leaf node. Corn plants produce ears at several consecutive leaf nodes, but the one located uppermost on the stalk usually develops to bear yield. This separation of the male and female reproductive parts dictates corn to reproduce mainly through cross-pollination within or between plants.

Sorghum, like most grasses, has an incomplete but perfect flower, with both its male and female parts located at the head (Figure 2.1). It has a panicle inflorescence in which the floral units are on a peduncle located above the flag leaf. Because the stamen and pistil are located at same position, sorghum is considered self-pollinating, although it may show 0–50% cross-pollination depending on genotype (Osuna-Ortega et al., 2003).

The differences in reproductive structure and pollination between sorghum and corn imply differences in drought tolerance and ease of plant breeding. Drought affects pollen viability through desiccation or by disrupting pollen shading and silk emergence time in corn more than it affects sorghum due to the distance between the two floral structures. On the other hand, hybrid production through breeding is easier for corn than sorghum due to the natural separation of the two floral structures.

2.2 PHYSIOLOGY

2.2.1 Photosynthetic Rate

Sorghum and corn are C_4 plants; therefore, they have a more efficient photosynthesis system than C_3 species (Zelitch, 1971). Under full sunlight and nonstressful conditions, a photosynthetic rate of about 60 mg CO_2 dm^{-2} h^{-1} was reported for both corn and sorghum (Singh et al., 1974). In the same article, Singh et al. (1974) reported photosynthetic rates of 177 and 182 mg CO_2 (g dw hr)$^{-1}$ for sorghum and corn, respectively, on a leaf-dry-weight basis under full sunlight. A reduction in available oxygen did not affect photosynthetic rates of either crop. Similar photosynthetic rates of 44 and 45 mg CO_2 dm^{-2} h^{-1} were reported for corn and sorghum, respectively, under oxygen (O_2)-limited (1–3% O_2) and in normal air (21% O_2) conditions (Hesketh, 1967). No direct effect of elevated CO_2 on photosynthetic rate in C_4 plants has been reported (Ghannoum et al., 2000; Kim et al., 2006; Leakey, 2009).

Unlike the minimal effects of CO_2 enrichment or reduced available O_2 discussed above, water stress significantly affects photosynthetic rates in C_4 plants (Ghannoum et al., 2000), primarily because of (i) stomatal closure and the consequent CO_2 limitation and (ii) other non-stomatal factors such as reduced enzymatic activity due to water status at the cellular level. Sorghum and corn respond differently to water stress; therefore, their photosynthetic rates vary at different levels of water stress. Field research has shown that sorghum can maintain a higher photosynthesis rate at lower water levels than corn (Turner, 1974; Ackerson and Kreig, 1977; Beadle et al., 1973; Boyer, 1970). Sorghum's advantage over corn in extracting water from deep in the soil profile and removing most of the apparent available water along with its ability to maintain stomatal openings at low levels of water potential are among reasons cited for these differences.

Relatively high differences in photosynthetic rates do not necessarily imply that sorghum maintains the same photosynthetic rate in normal and stress conditions. Water stress before and after flowering reduced the photosynthetic rate by about 37% (from about 58–36 mg CO_2 $dm^{-2} h^{-1}$) and 24% (from about 52–39 mg CO_2 $dm^{-2} h^{-1}$) in sorghum (Sung and Krieg, 1979). Up to a 60% decline in photosynthetic rate due to water stress at the tasseling stage was reported for corn (Atteya, 2003). Because similar photosynthetic rates were reported for the two crops under nonstress conditions, a 60% decline in photosynthesis means corn photosynthesis is lower than that of sorghum.

2.2.2 Transpiration and Water Uptake

Crop water requirements are a function of crop characteristics, management, and environmental demands. About 600–650 mm of water was required for a complete corn growing season at Colby, KS (Lamm et al., 2009). For maximum productivity, depending on hybrid maturity type and environment, about 450–650 mm of water was suggested for sorghum (Assefa et al., 2010; Lemaire and Hébert, 1996). The maximum yield of sorghum reported for different locations in Texas was achieved with evapotranspiration from 535 to 628 mm (Tolk and Howell, 2008). In another independent study in Texas, seasonal evapotranspiration recorded for maximum irrigated corn yield was 667–789 mm (Musick and Dusek, 1980). A comparative study of water use by three crops reported by Howell et al. (1994) summarizes an average of 578 and 771 mm ET for sorghum and corn, respectively. These research results show a relatively greater water use for corn than sorghum for maximum productivity.

A yield and water relationship curve by Stone and Schlegel (2006) for Tribune, KS, showed that the maximum yield of dryland sorghum (~ 8 Mg ha^{-1}) can be obtained from 300 mm of soil water at the beginning of growing season and an additional 300 mm of water via precipitation from June through September. Choosing corn over sorghum or vice versa depending on available water might depend on different factors. If yield were the only deciding factor, a comparison of grain sorghum and corn by Stone et al. (1996) showed that, for available irrigation water greater than 206 mm (8 in.), corn outyields sorghum; however, grain sorghum is a better choice if available irrigation water is less than 206 mm. The normal average annual precipitation for Tribune, KS, where the above conclusion was made, is about 17 in. (431 mm), and the soil is deep silt loam Ulyssess (fine-silty, mixed, mesic Aridic Haplustoll) with good water-holding capacity (Stone et al., 1994).

2.2.3 Cold-Temperature Tolerance

Corn and sorghum are both warm-season crops, but their seedlings have different degrees of cold tolerance. Much effort has been devoted to corn cold-tolerance research (Mock and Bakri, 1976; McConnell and Gardner, 1979; Mock and Eberhart, 1972). Corn has a broader range of temperature adaptation than sorghum. Nonfreezing low temperatures of 10–15°C result in a slow emergence rate, reduced emergence percentage, reduced growth rate after emergence, and poor seedling establishment in sorghum (Yu et al., 2004).

2.2.4 Response to Population Density

Corn responds to a greater degree than sorghum to changes in planting densities (Stanger and Lauer, 2006; Norwood, 2001; Blumenthal et al., 2003). A quadratic plateau regression model that specified available resources and environmental conditions was suggested by different authors to establish the relationship between corn yield and planting density. This characteristic of corn, being unable to adjust yield for varying planting density for possible occasionally inadequate resources, disables corn's stability across environments (Tokatlidis, 2013). On the other hand, sorghum demonstrates adjustments to available resources through tillers and is, therefore, not as responsive as corn to variation in planting densities. A stable yield was obtained in wide-ranging plant population densities for sorghum (Wade and Douglas, 1990; Conley et al., 2005).

2.3 PHENOLOGY

Growth and developmental stages for corn and sorghum as described by Ritchie et al. (1996), Abendroth et al. (2011), and Vanderlip (1993) are presented in Table 2.2. An approximate number of days for each developmental stage are also indicated. Different terminologies are used to describe the growth and development of the two crops by the original authors; however, they refer to the same or similar stages or processes. A primary difference between the crops or hybrids within each crop is the length of time or heat units required to complete each developmental stage. Depending on relative maturity, corn requires from 2,100 (short season) to 3,200 (full season) growing degree days

Table 2.2 Grain Sorghum and Corn Growth and Developmental Stages

Grain Sorghum (Vanderlip, 1993)		Corn (Abendroth et al., 2011; Ritchie et al., 1993)	
Descriptive Terms for Growth Stage of a Medium-Maturing Sorghum Hybrid	Average Number of Days After Emergence	Descriptive Terms for Growth Stage of Corn	Days After Emergence
0—Emergence (VE)	0	VE—Emergence	0
1—Collar of third leaf visible (V3)	10	V3—Third leaf collar	9–12
2—Collar of fifth leaf visible (V5)	20	V5—Fifth leaf collar	14–21
3—Growing point differentiation (V7-V10)	30	V6—Growing point differentiation	21–25
4—Flag leaf stage	40	V(n)—nth leaf collar	
5—Boot stage	50	–	
Heading stage[a]	55	VT—Tassling	
6—Half bloom	60	R1—Silking	55–66
		R2—Blister	67–78
7—Soft dough	70	R3—Milk	75–98
8—Hard dough	80	R4—Dough	81–92
		R5—Dent	91–102
9—Physiological maturity	90	R6—Physiological maturity	110–121

[a]Stage that is not part of sorghum growth and development in Vanderlip (1993) but are parts of the developmental process.

(Neild and Newman, 1987), whereas sorghum requires 2,673 (short season) to 3,360 (full season) growing degree days (Kelley, 2006).

One basic and important difference between the growth and development of the two crops is their growth response in water stress conditions. Sorghum has been reported to have more developmental plasticity than corn. Sorghum avoids effects of moisture stress at critical stages by delaying or hastening development (Fischer et al., 1982; Whiteman and Wilson, 1965). In response to water stress early in the vegetative stage, sorghum delays its growth; when recovered, sorghum has the ability to compensate yield by producing tillers. If water stress occurs late in the growth stage, sorghum hastens its growth and quickly passes to the next developmental stage.

A review of relevant scientific literature shows that morphologically, grain sorghum has deeper and denser roots, with a greater percentage of these roots in the topsoil than corn. Corn tends to be taller and has a greater number of leaves than sorghum. Physiologically, both crops have a similar photosynthetic rate and differ little in water requirements for maximum yield; however, in water stress conditions, sorghum maintains relatively higher photosynthetic rates and can yield better than corn. Similar growth and developmental stages have been documented for both crops, but different terminologies are used for each. Sorghum requires more heat units to complete its growth than corn when crops are compared within their maturity groups. The literature also suggests that sorghum has more developmental plasticity than corn in water stress conditions.

From the results of this review, we can clearly conclude that in an optimal situation in which water is not limiting, the crops have similar physiological and developmental characteristics; however, corn tends to produce more leaves, captures more radiation, and can convert that into superior yield compared with sorghum. Therefore, if yield is the only factor under consideration in optimal growing conditions, corn is the better choice. When water is limiting, sorghum possesses several morphological, physiological, and phenological drought tolerance and avoidance characteristics that make it superior to corn.

In the review process, we noticed that most morphological, physiological, and phenological characterizations of corn and sorghum were conducted in the early and mid-1900s. A number of factors have changed since then: (i) a number of drought-tolerant hybrids have been released for both crops, (ii) definitions of morphological parts of crops have been revised, and (iii) a number of new and accurate techniques have been devised to study these characteristics (e.g., root scanners for root architectural studies and portable photosynthesis measuring tools). Therefore, future research should prioritize creating a clearer picture of the differences in morphology, physiology, and phenology of the two crops.

CHAPTER 3

Corn and Grain Sorghum Historical Yield, Area, and Price Trends

The total harvested area of corn in the United States was about 40 million hectares from 1900 to 1930. This area started to decrease in the 1930s and was at its lowest in the 1960s, with an average of 23 million hectares from 1961 to 1969. Since the early 1970s, however, corn area has been increasing; its average for 2000 through 2011 was about 30 million hectares (NASS, 2011). Sorghum harvested area in the United States, on the other hand, increased from an average of 1.6 million hectares from 1930 to 1939 to an average of 5.5 million hectares in 1960 through 1970 but has declined since the early 1970s (USDA, 2011; Mason et al., 2008).

Researchers have offered several reasons for the decline in sorghum and increase in corn production since the early 1970s. The widening yield gap between the two crops (Mason et al., 2008), the release of new hybrids of corn tolerant to water deficit and heat stress (Staggenborg et al., 2008), the perception that corn has better nutritional characteristics and is better suited to livestock nutritional requirements (Duch-Carvallo and Malaga, 2009), the better price of corn (NASS, 2007), the greater number of pest control options for corn than for sorghum (Kershner et al., 2012), and the allelopathic effect of sorghum following wheat and corn (Roth et al., 2000; Schmidt and Frey, 1988) have been cited as potential drivers of the diverging production trends for the two crops.

Mason et al. (2008) concluded that the rate of yield increase was three times faster for corn than sorghum in Nebraska. In addition, they documented that across all production environments and years where they conducted their experiment, corn produced 1.7–4.3 Mg ha^{-1} greater yield than sorghum. On the other hand, Staggenborg et al. (2008) documented greater corn yields in years with timely and adequate rainfall and greater sorghum yields in drought years (Norwood, 1999; Gordon and Staggenborg, 2003).

The State of Kansas provides a wide range of weather conditions from east to west and south to north. Long-term corn and sorghum production areas, yield, and price in Kansas can be a suitable platform to study trends in corn and sorghum production. The objective of this section was to investigate, compare, and document historical trends of grain sorghum and corn yield, area, and price in Kansas for 1866–2011. We used the historical survey yield record of USDA National Agricultural and Statistics Service to meet this objective.

3.1 ANALYTICAL PROCEDURES

Kansas is divided into nine agricultural statistical districts (Figure 3.1) by the USDA National Agricultural Statistics Service to facilitate agricultural data reporting. The 30-year average (1979–2009) maximum and minimum monthly temperatures and total monthly rainfall for these nine districts are also presented in Figure 3.1. Annual total rainfall decreases significantly from east to west and varies minimally from north to south. Greater variation exists in average monthly minimum temperatures compared with monthly maximum temperatures. The eastern part of the state has relatively higher average monthly minimum temperatures than the western part of the state.

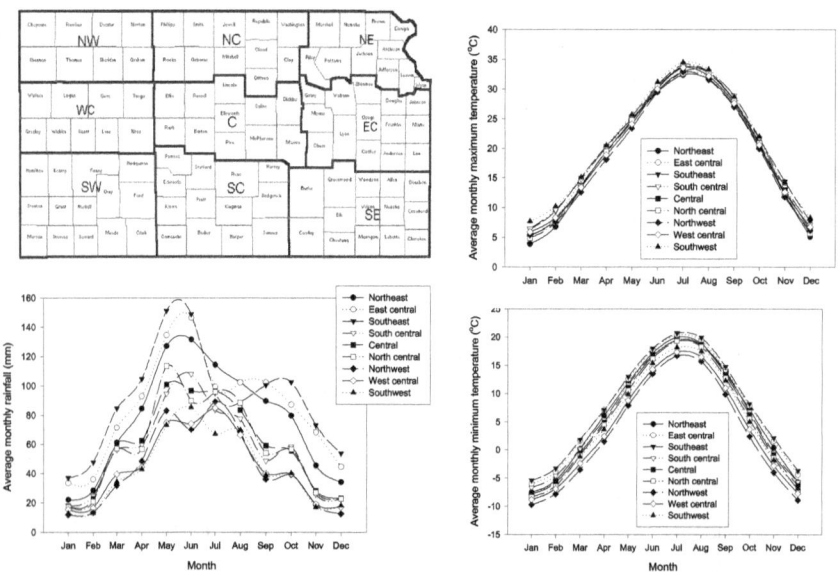

Figure 3.1 The nine agricultural statistics districts in Kansas as used by USDA and their 30-year average maximum and minimum monthly temperatures and total monthly rainfall.

In USDA National Agricultural Statistical Service historical records, the annual total harvest area and yield for corn in Kansas are available for 1866–2011 (USDA, 2011). The annual sorghum harvest area and yield for Kansas, however, are available only for 1929–2011. Corn and sorghum area and yield were plotted for the periods with data available for both crops in a time series plot using R statistical software version 2.15 for an exploratory analysis. Harvest area and yield stratified by irrigated and dryland production in Kansas were available only for years after 1972 for corn and after 1970 for sorghum; therefore, a comparison of harvested area and yield of corn in the nine crop reporting districts in Kansas by dryland and irrigated categories was conducted for 1970–2011.

In addition to the general exploratory trend analysis described, formal statistical tests comparing decade averages were also conducted. Harvested area and yield data for counties at each district in the dryland and irrigated system for the years 1970–2011 were used for these comparisons. The 1970–2011 period was divided into four decades, and the four crop classifications (irrigated and dryland corn and sorghum) were compared by crop reporting districts on a decadal basis. The four decades were 1970–1979, 1980–1989, 1990–1999, and 2000–2011. A test of similarity of means was conducted using PROC MIXED in SAS (SAS Institute, Inc., Cary, NC), and coefficients of variation were calculated by dividing the standard deviation into mean yields to compare variability between the two crops. Whenever necessary, a mean separation test was done using Tukey's Honest Significant Difference.

Corn and sorghum price data were available at the state level from the USDA National Agricultural Statistics Service. The monthly price of corn and sorghum for the years 1949–2011 was obtained and plotted. For formal analysis of changes in prices and their implications, the annual average price for the years 1949–2011 was divided into six decades: 1949–1959, 1960–1969, 1970–1979, 1980–1989, 1990–1999, and 2000–2011.

3.2 HARVESTED AREA

3.2.1 Trend of Corn and Sorghum Total Harvested Area

Historical trends of corn and sorghum total harvested area in Kansas are presented in Figure 3.2. Corn area was about five to six

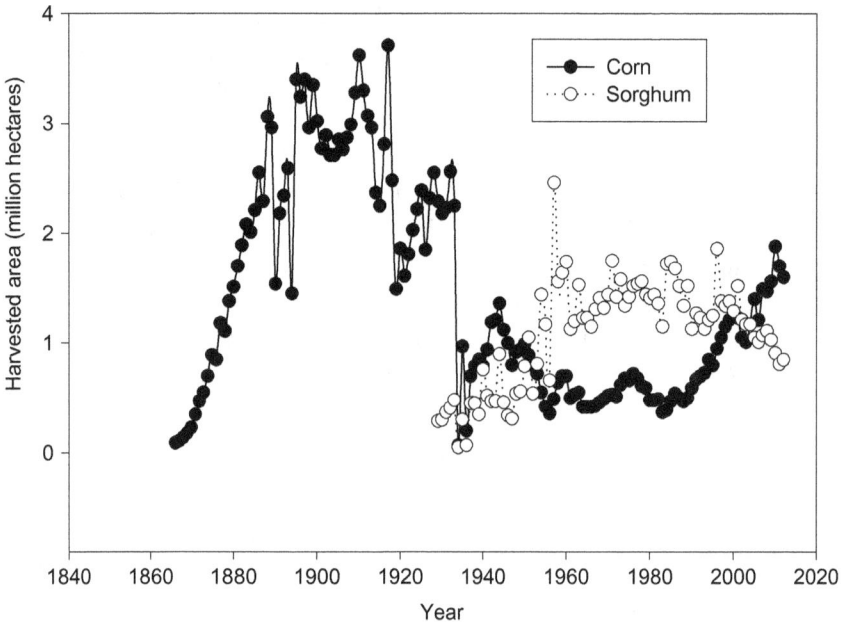

Figure 3.2 Time series plot of corn and sorghum total harvested areas in Kansas from 1866 to 2011.

times greater in the early 1900s than in the mid-1900s. A significant drop in harvested area of corn occurred in the 1930s, most likely due to the historic drought associated with the Dust Bowl years (Hornbeck, 2012). Corn maintained a greater harvested area than sorghum in most of the years from 1929 to 1950. Sorghum harvest area increased and surpassed corn for most of the years from 1950 to 2005 in Kansas, but corn harvested area increased by 48,000 ha yr^{-1} beginning around 1995 and has surpassed sorghum harvest area since 2005.

3.2.2 Trends of Corn and Sorghum Area in Irrigated and Dryland Cropping Systems

Figure 3.3 depicts the harvested area of dryland and irrigated corn and sorghum from 1970 to 2012 in Kansas. Harvested area of dryland sorghum was greater by far than irrigated sorghum or dryland and irrigated corn areas for the years 1970–2000; however, dryland corn harvested area started to increase in the mid-1990s, surpassed irrigated corn harvested area at the early 2000s, and has equaled or surpassed total sorghum harvested area since 2009.

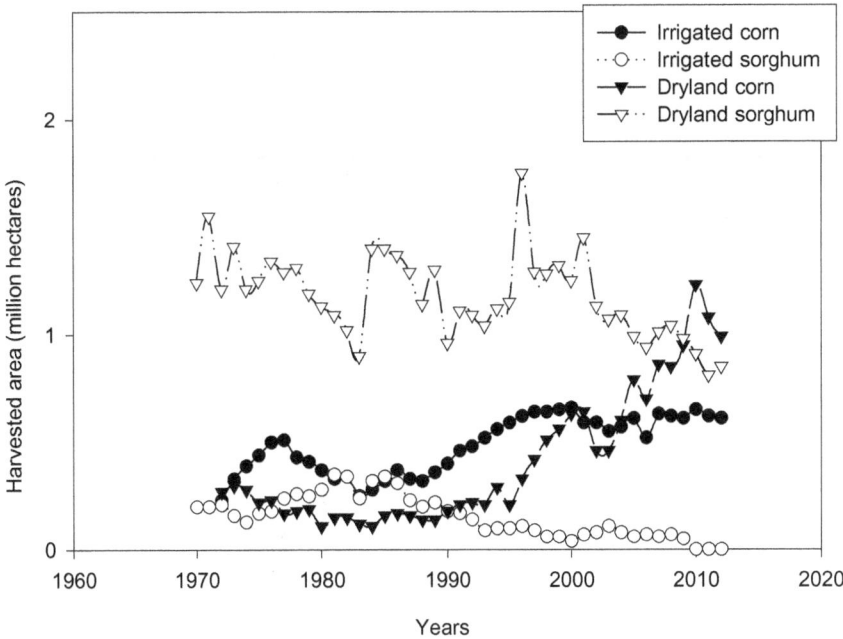

Figure 3.3 Time series plot of corn and sorghum harvested areas in Kansas from 1970 to 2011.

3.2.3 Trend of Harvested Area by District and Irrigated and Dryland Cropping Systems

In northwest Kansas, irrigated corn and dryland sorghum had greater harvested area compared with dryland corn and irrigated sorghum for most years between 1970 and the early 2000s (Figure 3.4). Since 1995, however, dryland corn harvested area has increased and surpassed both irrigated corn and dryland sorghum harvested areas by the early 2000s (Table 3.1). Dryland sorghum dominated harvested areas in most years from 1980 to 2005 in west-central Kansas, but dryland corn harvested area has increased since the early 2000s. In the southwest, irrigated corn and dryland and irrigated sorghum areas dominated from 1970 to 1980; however, irrigated sorghum area has declined, whereas both irrigated corn and dryland sorghum areas have continued to dominate. Irrigated corn area has plateaued since 2000, but dryland sorghum area has been highly variable in the south-central and southwest regions.

In central (north-central, central, and south-central) Kansas from 1970 to 2012, dryland sorghum area dominated compared with the areas for irrigated sorghum and dryland and irrigated corn, except in

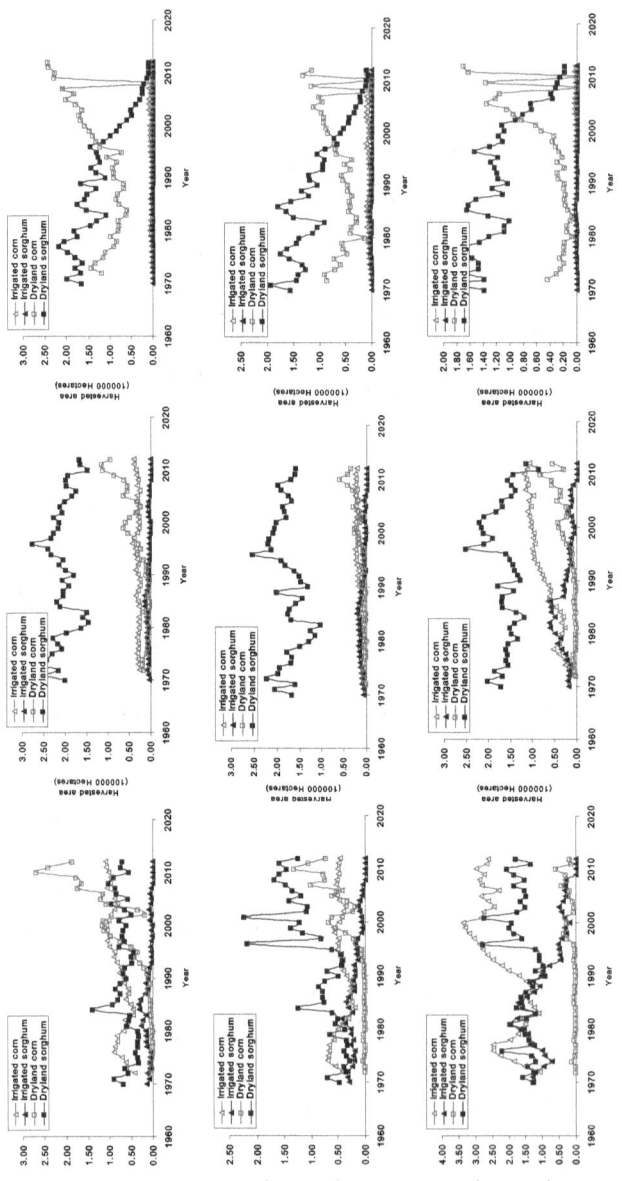

Figure 3.4 Time series plots of irrigated and dryland corn and grain sorghum harvest areas for each crop reporting district in Kansas from 1970 to 2011. Top panel is north, bottom panel is south, right panel is east, left panel is west, and center panel is central Kansas.

Table 3.1 Average Dryland Sorghum and Corn Harvested Areas Before and After 1995 and Rate of Change in Both Dryland Sorghum and Corn for the Years 1995–2012

Dryland Crop	Years	NE	EC	SE	NC	C	SC	NW	WC	SW
		1,000 ha district^{-1} or change 1,000 ha yr^{-1}								
Sorghum	1970–1995	159	135	137	207	164	154	61	56	139
	1995–2012	50	40	73	201	193	174	73	142	183
	Δ 1995–2012	−6	−6	−4	0	2	1	1	5	3
Corn	1970–1995	90	49	20	10	4	4	9	2	2
	1995–2012	171	82	80	60	24	37	128	59	24
	Δ 1995–2012	5	2	4	3	1	2	7	3	1

the south-central district, where irrigated corn area has been expanding since 1980 (Figure 3.4). Even so, dryland corn acreage have been increasing in these districts since the mid-1990s and even more dramatically since the mid-2000s. In the entire eastern third (northeast, east-central, and southeast) of Kansas, dryland sorghum harvested area followed by dryland corn dominated from 1970 to 1995; however, dryland sorghum harvested area has been decreasing since 1995 and has been surpassed by expanding dryland corn area since the mid-1990s in the northeast and east-central areas and since the early 2000s in the southeast district.

Unlike the national level, where corn harvested area has been greater by far than sorghum harvested area at all times, central and eastern Kansas historically were dominated by dryland sorghum until the mid-1990s. In the largest portion of western Kansas, irrigated corn and dryland sorghum occupied similar areas until the mid-1990s and early 2000s, when dryland corn acreage began to increase substantially. Significant declines in harvest area of dryland sorghum and significant increases in dryland corn areas have been prevalent in almost every district of Kansas in recent decades.

3.3 GRAIN YIELD

3.3.1 Trends of Corn and Sorghum Yields Average Across Cropping Systems

The average trend of corn and sorghum yields since 1866 in Kansas is depicted in Figure 3.5. From roughly 1929 to 1950, corn and sorghum

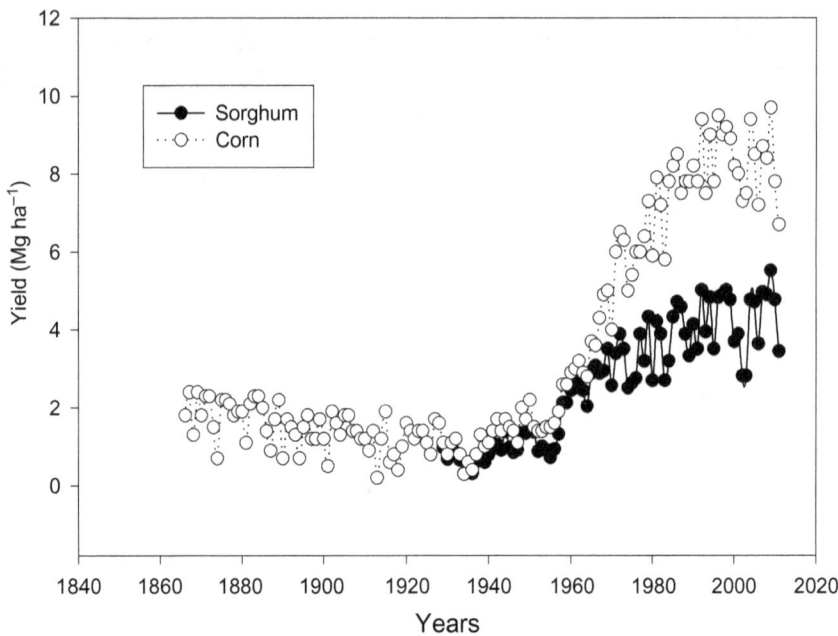

Figure 3.5 Time series plot of corn and sorghum yields in Kansas from 1866 to 2011.

had similar yields of about $0.7-1.5$ Mg ha^{-1} when averaged across irrigated and dryland production systems. The average yield difference between the two crops was less than 1 Mg ha^{-1} for the time between 1929 and 1960.

In the hybrid era, which began in the late 1930s for corn and late 1950s for sorghum, yield of both crops started to increase (Assefa and Staggenborg, 2010; Assefa et al., 2012). The average corn yield has been about $3-5$ Mg ha^{-1} greater than average sorghum yield since the 1970s. One of the main driving forces behind these differences in average yield is relatively high irrigated corn acreage compared with sorghum.

3.3.2 Corn and Sorghum Yield Trends in Irrigated and Dryland Cropping Systems

Irrigated corn has historically yielded more than the other three categories (dryland corn, irrigated, and dryland sorghum) and was the second most consistent after irrigated sorghum in terms of variability (Figure 3.6). For the 1970–2011 time period, about 15 kg ha^{-1} yr^{-1} yield gain was observed for irrigated corn. Dryland corn yields had the

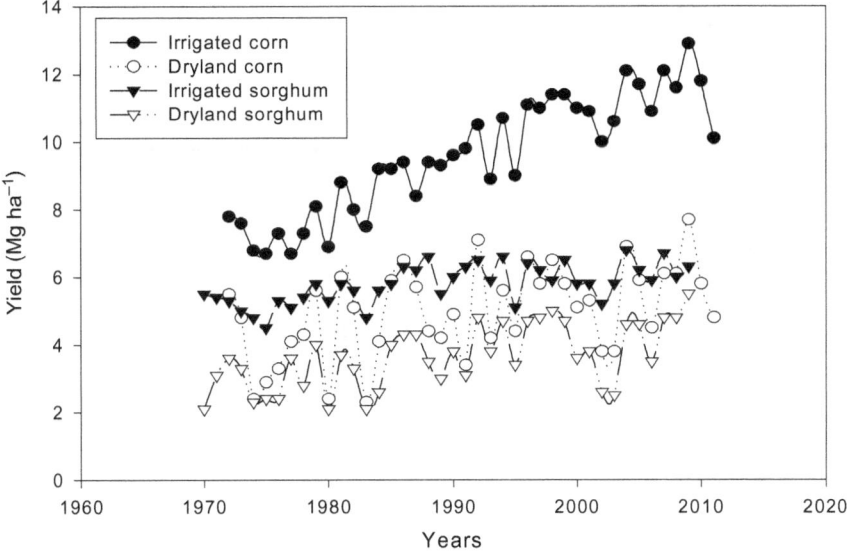

Figure 3.6 Time series plot of irrigated and dryland corn and sorghum yields in Kansas from 1970 to 2011.

highest year-to-year variability. The coefficient of variation (CV, an indicator of the variation in yield after standardizing for differences in mean values) calculated for the four decades from 1970 to 2011 for each category was 8.7, 24.7, 8.2, and 22.6 for irrigated corn, dryland corn, irrigated sorghum, and dryland sorghum, respectively. Greater coefficients of variation values in this situation indicate more year-to-year yield variability, whereas lesser values indicate more consistent yields from year to year.

3.3.3 Corn and Sorghum Yield Trends by Districts and Cropping System

The spatial and time series characteristics of corn and sorghum yields in Kansas are depicted in Figure 3.7. In western and central Kansas, similar to what was observed in the statewide average trends of the four variables above, irrigated corn was the highest yielding in all districts in all years from 1970 to 2011. Irrigated sorghum was the second highest yielding in the western and central Kansas regions, with yields clearly superior to both dryland corn and sorghum in most cases. In western and central Kansas, average dryland corn and sorghum had similar yields almost every year, but yields of dryland corn varied more from year to year.

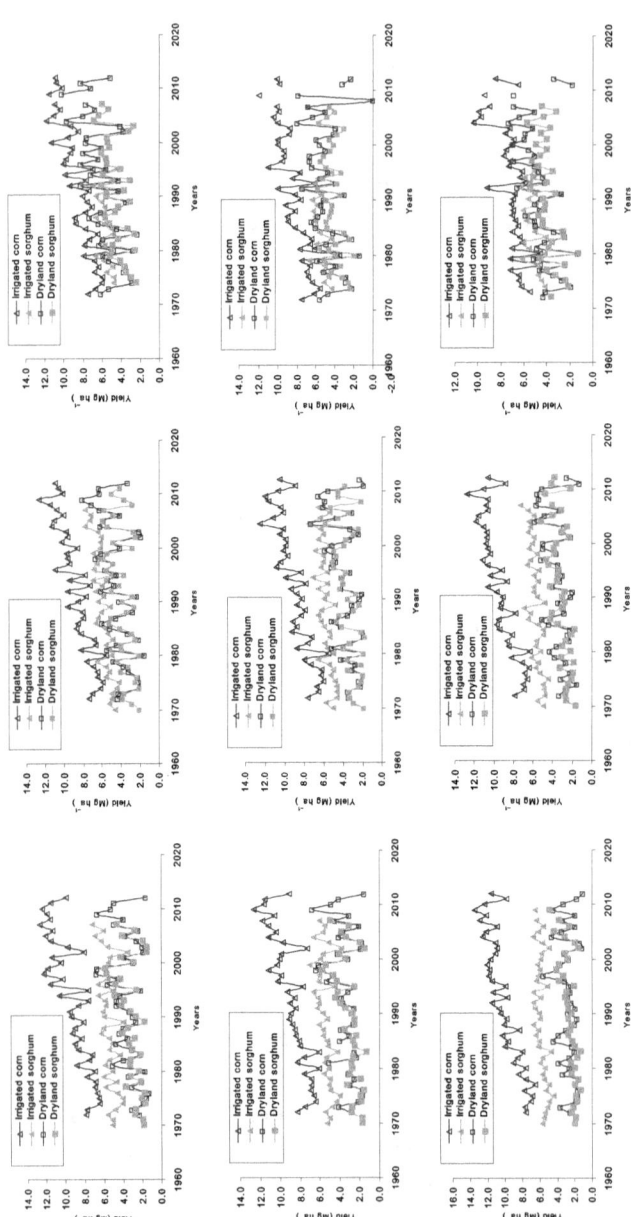

Figure 3.7 Time series plots of irrigated and dryland corn and grain sorghum yields for each crop reporting district in Kansas from 1970 to 2011. Top panel is north, bottom panel is south, right panel is east, left panel is west, and center panel is central Kansas.

In eastern Kansas, as in western and central Kansas, irrigated corn yields were clearly superior to irrigated sorghum or dryland corn or sorghum (Figure 3.7), but the gap between irrigated corn yields and the other categories was not as wide as that observed in the western and central districts. The average yield difference between irrigated sorghum and dryland corn or sorghum in eastern Kansas was not obvious in most years. Dryland corn and sorghum yields follow the same trend except in the last decade, when dryland corn yield was superior to dryland sorghum in most years in eastern Kansas (Figure 3.7). Eastern Kansas historically has annual rainfall above 800 mm, and an increase in rainfall in southeastern Kansas is projected and documented (Assefa et al., 2012). Therefore, the magnitude of water stress in dryland crops in central and western Kansas is different from eastern Kansas.

Table 3.2 presents a statistical comparison of dryland sorghum and corn yield in the nine crop reporting districts of Kansas for the four decades since 1970. No significant difference is apparent between dryland corn and sorghum in western and central Kansas in any decade except in southwest and south-central Kansas in 1980–1989, when dryland corn yielded more than dryland sorghum. The CV for dryland corn was greater than for dryland sorghum in all nine districts.

3.3.4 Relationships Between Dryland and Irrigated Corn and Sorghum Yields

Figure 3.8 shows the correlation between dryland and irrigated corn and sorghum yields in Kansas. Except for the correlation between dryland and irrigated sorghum, all correlations were positive and significant; however, a relatively strong correlation—close to one-to-one— was obtained only for dryland corn versus dryland sorghum.

From about 360 data points used to observe the relationship between irrigated corn and dryland corn, only two were on the one-to-one line. All the other data points fall below the one-to-one line, toward irrigated corn, showing the clear superiority of irrigated corn over dryland corn at all times and in all districts. The regression line between irrigated and dryland corn crosses the one-to-one line at about 1 Mg ha^{-1}, which indicates that dryland corn would be superior to irrigated corn only when yields of irrigated corn are less than 1 Mg ha^{-1}. Even so, this part

Table 3.2 Average Dryland Corn and Sorghum Yields in Nine Kansas Districts and Four Decades from 1970

Decade	Dryland Crop	NW	WC	SW	NC	C	SC	NE	EC	SE
						Average Yield, Mg ha^{-1}				
1970–1979	Corn	2.4cd†	2.6cd	2.3c	3.0e	2.7c	2.7cd	4.2de	4.0bc	3.6c
	Sorghum	2.4d	2.2d	–	3.0e	2.6c	2.3d	3.9e	3.6c	3.3c
1980–1989	Corn	3.3b	3.0bc	3.0ab	3.7cd	3.1bc	3.3b	4.7cd	4.6a	4.1b
	Sorghum	3.0bc	2.6cd	2.3c	3.5de	3.0bc	2.8c	4.2de	3.9bc	3.5c
1990–1999	Corn	4.5a	4.0a	3.2ab	4.9ab	4.0a	3.3b	5.4b	4.9a	4.3b
	Sorghum	4.1a	4.0a	3.2a	4.3bc	3.9a	3.4b	5.0bc	4.6a	4.4b
2000–2011	Corn	3.4b	3.5b	2.9ab	4.9a	4.4a	4.0a	6.4a	5.2a	6.0a
	Sorghum	2.9bcd	3.1bc	2.9b	4.5abc	3.7ab	3.5ab	5.1bc	4.3ab	4.4b
CV	Corn	37.6	36.7	38.9	35.0	37.2	35.0	34.9	35.3	31.7
	Sorghum	31.9	32.5	31.8	30.4	31.0	30.1	24.4	25.1	25.8

† Numbers in a column followed by same letters are not significantly different.
NW: northwest, WC: west-central, SW: southwest, NC: north-central, C: central, SC: south-central, NE: northeast, EC: east-central, SE: southeast

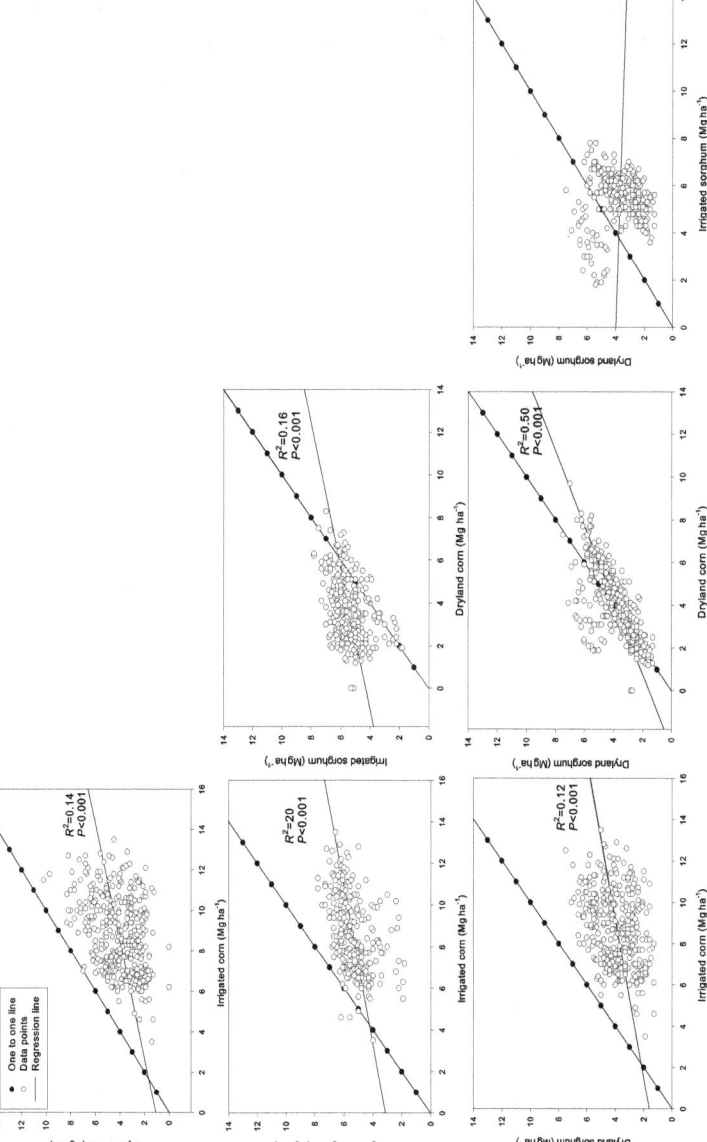

Figure 3.8 Relationships between irrigated and dryland corn and sorghum. The closer the open-dotted points to the one-to-one points, the more similar the yield between the two categories.

of the regression line is beyond the range of observed yield values, making predictions in this yield range of little value.

The regression line for the relationship between irrigated corn and irrigated sorghum crosses the one-to-one line at about 4 Mg ha^{-1}. This crossing point and the data points that are above the one-to-one line imply that irrigated sorghum could be superior to irrigated corn only when yields are at or below about 4–4.5 Mg ha^{-1}. Similarly, irrigated corn was superior to dryland sorghum at all yield levels; that is, there were no instances where dryland sorghum yielded equally or superior to irrigated corn.

Irrigated sorghum was superior to dryland corn in most instances, but a number of instances showed similar yields for the two systems (Figure 3.8). Based on the crossing points of regression lines, we can deduce that in locations where dryland corn yields are expected to be more than 6 Mg ha^{-1}, it is likely to yield more than irrigated sorghum.

Dryland corn and sorghum yields were the most similar of the four categories, exhibiting a relationship that was closer to one-to-one. With almost equal frequency, one was slightly superior to the other. In a number of instances, dryland sorghum yields were clearly superior and were far from the one-to-one line. The crossing point of the regression line with the one-to-one line implies that in environments where dryland corn yields are equal to or less than 4 Mg ha^{-1}, dryland sorghum yields were superior.

3.4 PRICES OF CORN AND SORGHUM, 1949–2011

A time series plot of corn and sorghum prices from 1949 to 2011 and decadal averages are presented in Figure 3.9. Prices of sorghum and corn followed a similar pattern. The price of both crops has been increasing over time, with the greatest price jump recorded in the most recent decade.

The corn price was almost always greater than the sorghum price, with a minimum average price difference of $0.007 kg^{-1} in 1949–1969 and a maximum price difference of about $0.014 kg^{-1} in 1980–2000. The price of both crops was relatively less in 1949–1973 and nearly doubled by about 1974. The average price difference between the two crops in 1974–2011 was also about twice the difference in 1949–1973.

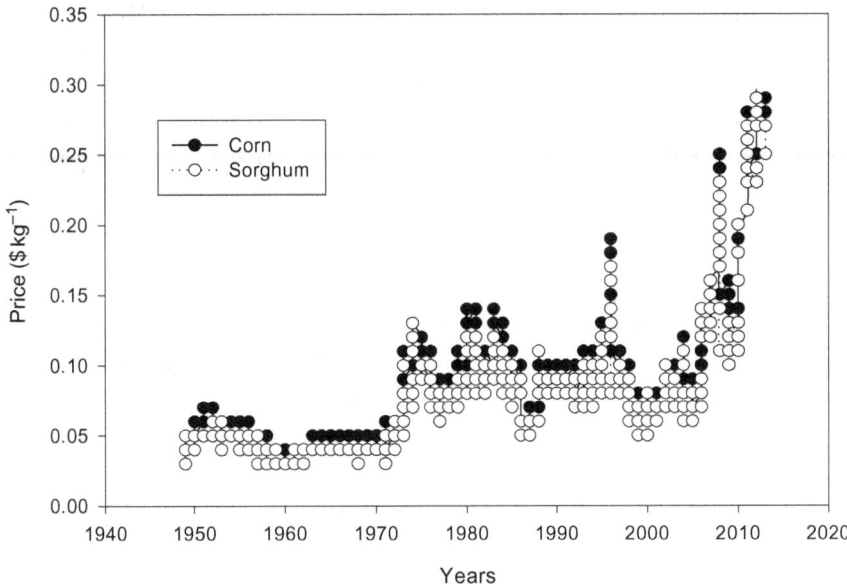

Figure 3.9 Time series plot of corn and sorghum prices in Kansas from 1949 to 2011.

The prices for both crops have reached new highs in the past few years, nearly triple the 1974–2006 average. The price differential between the two crops at these new highs is $0.013.

In this section, we documented increasing corn and declining sorghum harvest areas for Kansas, similar to what has been reported by many authors for the United States (Hamman et al., 2002; Mason et al., 2008; Staggenborg et al., 2008). The analysis breaks down the comparison between sorghum and corn into dryland and irrigated categories and into crop reporting districts. Results show that irrigated corn yields historically have been superior to dryland corn and irrigated or dryland sorghum in almost every district of Kansas. The gap between irrigated corn yields and other yields has increased in the study's time period, and this result is similar to the report by Assefa et al. (2012), which used data from Kansas Corn Performance Trials. Therefore, if yield were the only measure, irrigated corn is the most productive.

Comparisons of dryland corn and sorghum yields indicate that dryland corn yields have been superior in the eastern part of Kansas in recent decades. This superiority in the east is likely the result of historically wetter conditions in that part of Kansas, a difference that has become more pronounced in recent years (Assefa et al., 2012). In central and western Kansas, dryland corn and sorghum yields did not show a significant difference in most years. The higher CV noted for dryland corn indicates that it is less stable than dryland sorghum in these environments.

Yield cutoff values for corn and sorghum will be discussed in detail in Chapter 6. Assuming USDA yield data for the two crops coming from adjacent fields and assuming yield were the only decision factor, equivalent yield points derived from correlation results (Figure 3.9) can be used to determine cropping decisions for irrigated and dryland corn and sorghum. In both dryland and irrigated conditions, an expected yield of 4–4.5 Mg ha^{-1} can be used as the cutoff point. For locations or in years where expected yields are about 4–4.5 Mg ha^{-1} or less, dryland sorghum might be the best choice, but corn will be the best choice with greater expected yields.

CHAPTER 4

Corn and Grain Sorghum Yield Trend Since the Beginning of Hybrid Technology

4.1 INTRODUCTION

Mean corn and grain sorghum yields in the United States have improved from about 1 Mg ha^{-1} in the early 1900s to more than 8.5 and 3.5 Mg ha^{-1} in the early twenty-first century, respectively (USDA National Agricultural Statistical Service, 2009). For corn, yield gain of about 110 kg ha^{-1} yr^{-1} was estimated for the years from 1950 to 1980 (Duvick, 1984; Castleberry et al., 1984). Eghball and Power (1995) reported a sixfold yield increase for corn from 1930 to 1990. Troyer (2000) estimated that the average corn yield gain was 63 kg ha^{-1} yr^{-1} from 1930 to 1960 and 110 kg ha^{-1} yr^{-1} from 1960 to 2000. From an analysis of 61 years (1930–1990) of sorghum yield data, Eghball and Power (1995) reported a 50 kg ha^{-1} yr^{-1} yield increase in the United States. Similarly, based on studies conducted at the USDA-ARS Conservation and Production Research Laboratory, Bushland, TX, Unger and Baumhardt (1999) reported that sorghum grain yield increased by about 139% (1600–3800 kg ha^{-1}) from 1956 to 1997.

Various management, weather, and genetic factors contribute to yield increases. For corn, genetic-attributable yield increase was estimated by different authors as 33% (Hallauer, 1973), 63% (Russell, 1974), and 57% (Duvick, 1977) for the years from 1930 to 1980. Advances in crop management practices, such as improvements in harvest equipment, changes in planting date, use of inorganic fertilizers, enhanced weed and pest control, and increased planting density, contribute a significant portion of yield gain (Cardwell, 1982; Edmeades and Tollenaar, 1990). Duvick (2005) summarized corn yield changes and contributions to those changes and concluded that on an average, 50% of the yield increase in corn was due to changes in management and the other 50% was due to genetics. According to Duvick (1999), approximately 35–40% of the total yield gain in grain sorghum is assumed to be due to hybrid improvement. Changes in cultural practices, like N fertilizer, irrigation, and tillage, were assumed to be responsible for 60–65% of the yield gain. Unger and Baumhardt (1999) indicated that 1.5 of the 2.2 Mg ha^{-1} yield increase they reported was a result of increased soil water at planting due to changes in management practices; the other 0.7 Mg ha^{-1} was attributed to improved hybrids.

The relationship between changes in climate and crop yield trends also has received attention recently. Based on their analysis of county-level USDA corn yield data from 1982 to 1998, Lobell and Asner (2003) reported that about 25% of corn yield increase was due to increased temperature. They concluded that previous estimates of corn yield increase attributed to technology might have been overestimated by about 20%. Kucharik and Serbin (2008) also reported that 40% of the corn yield trend from 1976 to 2006 in Wisconsin was due to changes in precipitation and temperature. The positive impact of the projected climate change scenarios from the Intergovernmental Panel of Climate Change (Metz et al., 2007) on corn yield in the United States, Canada, and China was also recently reported (Li et al., 2011). On the other hand, Tannura et al. (2008) concluded that there was no upward or downward trend in temperature and precipitation for the years 1960 through 2006 at sensitive stages of corn development for Illinois, Indiana, and Iowa. Hu and Buyanovsky (2003) did not find a trend in precipitation and temperature for years 1895 through 1998 for Missouri nor did Assefa and Staggenborg (2010) for the years 1957 through 2008 in Kansas for sorghum growing seasons.

Almost all available corn and grain sorghum yield trend analysis reports are based on USDA data, which are collected through field observation and survey. Most of these reports lack a separate comparison for irrigated and dryland yield trends. From their study of irrigated and dryland corn yields, Duvick and Cassman (1999) documented small changes in irrigated corn yields but major yield increases in dryland corn yields; however, this comparison between irrigated and dryland corn yields was done only for two decades (1983–2000). A detailed analysis of irrigated and dryland corn and sorghum yield for the full length of the hybrid era, which is 1930s to the present for corn and 1957 to the present for sorghum, based on measured yield data rather than yield survey estimates was necessary.

Multilocation corn and sorghum hybrid trials have been conducted in Kansas (Clapp et al., 1939; Kansas State University, 1939–2009;

Kansas grain sorghum performance trial reports, 1957–2008). These trials have included a number of improved hybrids that changed as new hybrids were released. They are consistent research platforms that can offer insight into yield trends that span several decades across a broad geographic area in Kansas.

The main objectives of this part of the study were to determine the magnitude of yield changes in irrigated and dryland corn and sorghum in their respective hybrid era, to indicate potential factors that contributed to those yield changes and to compare this crop on these bases (yield changes and contributing factors).

4.2 CORN FROM 1939 TO 2009

4.2.1 Data Organization and Analytical Methods for Corn

Corn hybrid performance trials have been conducted in Kansas since 1926 (Clapp et al., 1939), and trial data are available since 1939 (Kansas State University, 1939–2009). Management practices for the yield trials evolved to match those used by farmers, and the trials have separate irrigated and dryland components. The experimental design and the number of replications varied very little in time, locations, and between irrigated and dryland trials. Corn hybrids were planted in small plots of 9.3–27.9 m^2 and replicated three or four times in a randomized complete block design. Hybrids were planted twice the desired density and hand thinned from 1939 to 1995. From 1995 to 2009, hybrids were planted at 10–20% more than the desired density depending on expected emergence rate for each trial. This change was facilitated by improvement in planting equipment and caused no significant increase in variability of yield estimates (Roozeboom et al., 1994). Hybrids of corn plant, management practices, and climate, however, varied in time, location, and between irrigated and dryland trials and they are core of the analysis.

Data from these dryland and irrigated corn performance trials conducted in each of eight regions in Kansas (Figure 4.1) were analyzed. Regions were classified based on average annual rainfall as follows: eastern Kansas (northeast, east-central, southeast), above 800 mm rain yr^{-1}; central Kansas (north-central and south-central), from 550 to 800 mm rain yr^{-1}; and western Kansas (northwest, west-central, and southwest), below 550 mm rain yr^{-1}. Table 4.1 shows Kansas

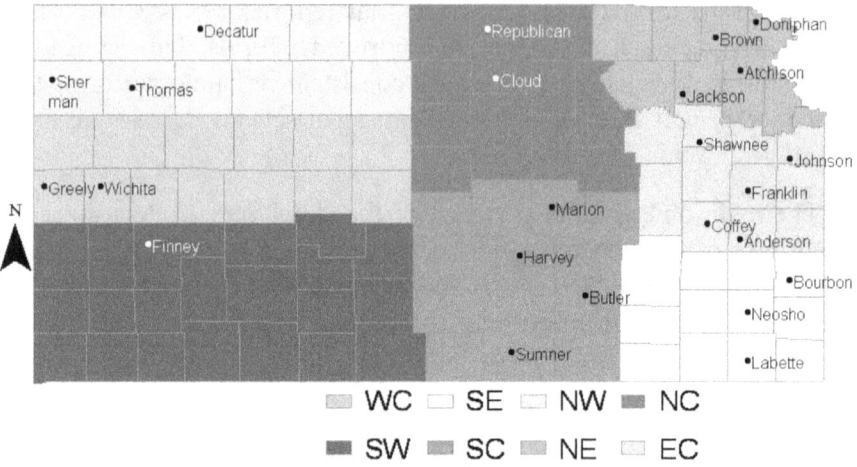

Figure 4.1 Map of the state of Kansas depicting regions and counties where corn hybrid performance trial and climatic data were collected and analyzed.

Table 4.1 Kansas Corn Hybrid Performance Trial Site Location in Counties, Annual Precipitation, and Years for Which Data Were Analyzed

Major Areas	Region	Trial Type	Trial Sites Location (Counties)	Annual Precipitation (mm)	Yield, Management, and Climatic Data Available for Years	
					Earliest	Latest
East Kansas	Northeast	Dryland	Atchison, Brown, Jackson, Doniphan	800–900	1939	2009
	East-central	Dryland	Franklin, Shawnee, Coffee, Anderson, Johnson	800–1,000	1939	2009
		Irrigated	Shawnee		1959	2009
	Southeast	Dryland	Bourbon, Neosho, Labette	900–1,000	1940	2009
Central Kansas	North-central	Dryland	Cloud, Republican	550–800	1939	2009
		Irrigated	Cloud, Republican		1954	2009
	South-central	Dryland	Butler, Harvey, Marion, Sumner	550–800	1939	2009
		Irrigated	Stafford		1969	2009
Western Kansas	Northwest	Dryland	Decatur, Thomas	< 550	1943	2009
		Irrigated	Sherman, Thomas		1958	2009
	West-central	Irrigated	Greely, Wichita		1960	2009
	Southwest	Irrigated	Finney		1962	2009

corn performance trial locations, precipitation ranges of regions, and years of data used from each location.

Two yield data sets were assembled for each trial site: average yield of all hybrids tested and yield of the top-performing hybrid. In addition, management data for each year including planting date, harvest date, planting density, and NPK fertilizer applications were collected. Weather data were collected from the Kansas State University Weather Data Library for each year and each site where yield data were available.

Average corn hybrid yield data were compared by regions and years. This comparison was conducted for irrigated and dryland locations, separately, using PROC GLM in SAS (SAS Institute Inc., Cary, NC, 2001) with the model equating yield with regions, years, and their interaction. Because the main effects of years and region were found to have a significant effect on yield but their interaction was not significant ($\alpha = 0.05$), only main effects were interpreted. To compare regions, a mean separation test was conducted using Duncan's multiple test ($\alpha = 0.05$). The average yield of hybrids tested over the 70 years was regressed against time for possible yield trends. This analysis was conducted using PROC REG in SAS, equating yield with years in the model statement.

To determine what contributes to change in dryland and irrigated corn yields through the years, analysis of crop management, soil management, climatic and genetic factors was conducted for the same time period. An initial analysis was conducted to test changes in planting date, harvest date, planting density, and NPK fertilizer applications through the years. A subsequent analysis was conducted to characterize potential relationships between yield and the management factors. These analyses were conducted using PROC GLM and PROC REG in SAS, equating these management factors with years and yield in the model statement.

The timing of significant impacts of each of the management factors on yield was determined using a decade-by-decade analysis of management factors and yield. Decades were designated 1–7 beginning in 1939; i.e., 1939–1948 was decade 1 and 1999–2009 was decade 7. Changes in decadal management factors were analyzed using PROC GLM in SAS, with model equating management factors with decadal class. A mean separation test between decadal averages of each management factor was conducted using Duncan's multiple tests ($\alpha = 0.05$).

Analysis of change in total rainfall and average maximum and minimum temperatures during the years and their correlation with yield was conducted using PROC CORR in SAS. The changes detected in these climate variables over the years were highly location- and time-dependent. Therefore, correlation analysis of weather trend was conducted for each location in each month.

Unlike the management and climatic factors considered herein, change in corn hybrids, or equivalently change in genetics, was not a continuous variable to be directly used for correlation analysis. Therefore, we have introduced the term "replacement percentage" to quantify and relate yield changes attributable to genetics. Replacement percentage is the number of new hybrids in the performance trials each year or each decade expressed as a percent of the total number of hybrids that year or that decade. As a general breeding practice, hybrids are released after several years of testing against the best commercial hybrids in the region they are about to be used. There is a desire to add at least 2–5% yield advantage with advanced hybrids while at the same time enhancing the needed disease resistance and agronomic traits that can change over time (Keaschall J., research director, Pioneer Hi-Bred Intl., 2011, personal communication). Therefore, a new hybrid usually carries a yield advantage purely from genetics and it was important to quantify the turnover of hybrids each year or decade in the performance trials.

Changes in genetics in Kansas Performance Trials were analyzed using the "replacement percentage" of hybrids entered every year and every decade using hybrids from Pioneer Hi-Bred, Intl., Inc. (Johnson, IA). Pioneer brand hybrids were used for replacement percentage analysis only because the company is one of the few hybrid companies that entered hybrids in the performance trial for the length of our analysis. The total number of Pioneer brand hybrids entered each year in corn performance trials from 1939 to 2009; the number of new hybrids in each year compared with the previous year; and the number of new hybrids each year from 1949 to 2009 compared with hybrids more than 10 years back were collected as part of the data set from performance reports. This information was used to calculate the replacement percentages that were used in the analysis. Calculating replacement percentage using hybrid from a single company prevented over estimation of hybrid turnover resulting from sporadic participation of some companies or from brand and hybrid name changes associated with seed company consolidation.

4.2.2 Results
4.2.2.1 Average Irrigated and Dryland Yield

Average dryland corn yield was different between regions and between years, but the interaction of regions and years was not significant (Table 4.2). Mean dryland yields of all hybrids tested from 1939 to 2009 ranged from 4.1 Mg ha^{-1} in south-central Kansas to 7.5 Mg ha^{-1} in northeast Kansas. The mean yield of the highest yielding corn hybrids tested in dryland from 1939 to 2009 ranged from 5.2 Mg ha^{-1} in south-central Kansas to 8.8 Mg ha^{-1} in northeast Kansas. Northeast Kansas had the greatest mean dryland yield. Both measures of hybrid yield decreased in regions farther to the south and west.

Table 4.2 Mean Yield of All Hybrids and the Highest Yielding Hybrids in Corn Hybrid Performance Trials Conducted in Different Locations of Kansas from 1939 to 2009

Region in Kansas	Dryland		Irrigated	
	Mean Yield of Hybrids (Mg ha^{-1})	Mean Yield of Highest Yielding Hybrid (Mg ha^{-1})	Mean Yield of Hybrids (Mg ha^{-1})	Mean Yield of Highest Yielding Hybrid (Mg ha^{-1})
Northeast	7.5a[†]	8.8a	–	–
East-central	6.5b	7.7b	10.5b	12.3b
Southeast	6.3b	7.6b	–	–
North-central	5.8b	6.9b	10.4b	12.1b
South-central	4.1c	5.2c	10.8b	12.7b
Northwest	4.3c	5.2c	11.8a	13.5a
West-central	–	–	11.0b	12.7b
Southwest	–	–	10.7b	12.6b
Years	***	***	*	*
Regions	***	***	*	*
Years × regions	ns	ns	ns	ns

*** Significantly different at P ≤ 0.001.
* Significantly different at P ≤ 0.05.
ns = not significantly different.
[†] Within columns, means followed by same letter are not significantly different at P ≤ 0.05.

Similarly, average irrigated corn yield was different between regions and between years, but the interaction of regions and years was not significant (Table 4.2). Both mean irrigated corn yield of all hybrids tested and the mean yield of highest yielding hybrid were superior in northwest Kansas, 11.8 and 13.5 Mg ha^{-1}, respectively; however, irrigated yields were not different between the rest of the regions. Irrigated corn had about a 70% yield advantage over dryland corn.

Regression analysis found a significant increase in both dryland and irrigated corn yield in all regions over the time period considered (Figure 4.2). On an average, yield increased by approximately

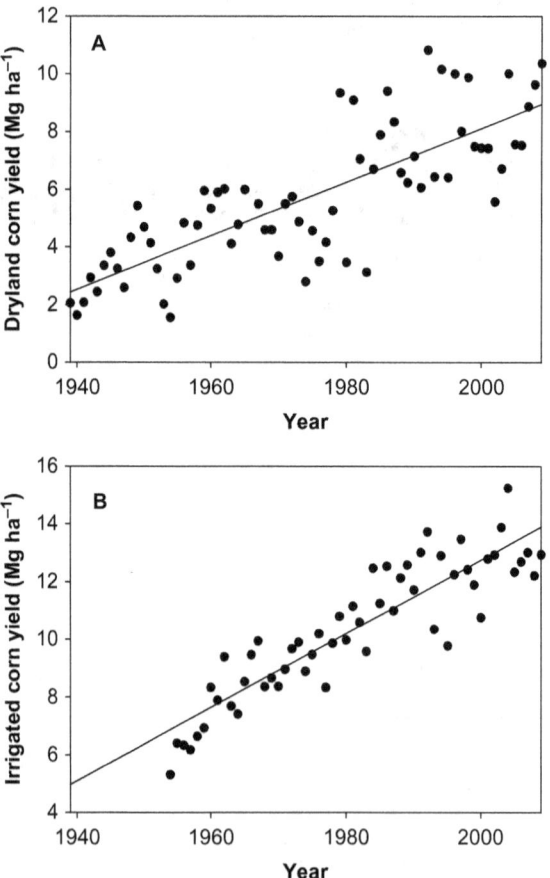

Figure 4.2 Average dryland corn yield from 1939 to 2009 (A) and average irrigated corn yield from 1954 to 2009 (B) from Kansas Corn Hybrid Performance Trial data.

90 kg ha^{-1} yr^{-1} in the dryland trial sites from 1939 to 2009 and by approximately 120 kg ha^{-1} yr^{-1} in irrigated trials from 1954 to 2009. Rate of change of yield for the dryland trials over the same time period for the irrigated trials, 1954 through 2009, was 86 kg ha^{-1} yr^{-1}, i.e., slightly lower than the rate of change of dryland yield for 1939 through 2009.

Our results agree with a better yield gain report for irrigated than for dryland corn hybrids released from 1953 to 2001 for central Iowa (Barker et al., 2005). Barker et al. (2005) reported a yield gain of 211 kg ha^{-1} yr^{-1} for irrigated and a 47–77% less yield gain for dryland (water stress) after comparing hybrids released in the years 1953 through 2001. The average rate of increase in corn yield reported here agrees with previous findings for other corn growing areas, even though most of the reports do not have separate irrigated and dryland trends. Troyer (2000) reported a 110 kg ha^{-1} yr^{-1} corn yield increase from 1960 to 2000 in the United States. A linear yield increase of 80 kg ha^{-1} yr^{-1} was reported for the years 1940 through 2000 for Canada (Bruulsema et al., 2000).

4.2.2.2 Management Factors that Contribute to Yield Changes

Significant ($\alpha = 0.05$) changes in planting and harvest dates, planting density, N, and fertilizer levels were found for the years 1939 through 2009 (Table 4.3). Planting and harvest date decreased by about a quarter of a day yr^{-1}; planting density increased by about 597 and 697 plants ha^{-1} yr^{-1}; N fertilizer levels increased by about 2.6 and 2.2 kg ha^{-1} yr^{-1}; P fertilizer levels increased by about 0.40 and 0.3 kg ha^{-1} yr^{-1}; and K levels decreased by about 0.14 and 0.06 kg ha^{-1} yr^{-1} in dryland and irrigated trials, respectively. The change in fertilizer application over the years was different from one district to another.

When management factors are evaluated independently, a significant correlation emerged between dryland yields with each of the factors. Yield correlates significantly with planting and harvest date, planting density, and NPK fertilizer applications (Table 4.4). Similarly, in irrigated trials, all management factors except harvest date were related to yield increases. The rate of change of yield for a unit change of each management factor, one management factor regressed independently, is given in Table 4.4.

Table 4.3 Changes in Planting and Harvest Date, Planting Density, and NPK Fertilizer Levels with Years and Their Relation to Corn Yield from Year 1 (1939) to Year 7 (2009)

Crop and Plant Management Factors	Intercept	Coefficient for Management for a Unit Change in Year	SE	R^2	Year × Region Interaction
Dryland Year					
Management Unit $(yr)^{-1}$					
Planting date (days from January 1)	131.4	−0.27	0.11	0.06	ns
Harvesting date (days from January 1)	285.3	−0.28	0.09	0.05	ns
Density (plant ha^{-1})	15,423	597	27.84	0.71	ns
N level (kg ha^{-1})	3.3	2.6	0.22	0.41	ns
P_2O_5 level (kg ha^{-1})	8.8	0.40	0.11	0.06	*
K_2O level (kg ha^{-1})	22.8	−0.14	0.11	0.01	*
Irrigated Year					
Management Unit $(yr)^{-1}$					
Planting date (days from January 1)	134.7	−0.27	0.15	0.04	ns
Harvesting date (days from January 1)	293.9	−0.25	0.05	0.07	ns
Density (plant ha^{-1})	27,235	697	0.76	23.0	ns
N level (kg ha^{-1})	104.9	2.2	0.13	0.32	*
P_2O_5 level (kg ha^{-1})	17.3	0.3	0.02	0.1	*
K_2O level (kg ha^{-1})	4.9	−0.06	0.01	0.03	*

ns = not significant at different $P \leq 0.05$.
*Significant at $P \leq 0.05$.

Decadal analysis of management factors indicated that planting date, on an average, becomes earlier from the first two decades through the fourth decade in both dryland and irrigated corn trials, but it has been stable since then (Figure 4.3). Harvest dates also varied, but changes in harvest dates were not smooth. Planting density, on the other hand, significantly increased from one decade to next in all seven decades from 1949 to 2009 in both irrigated and dryland corn.

Applied fertilizer rates, NPK, did not change from the first to the second decade, but a significant increase in the amount of NPK was found between the second and the fourth decade. The amount of N fertilizer stabilized from the fourth decade to the last decade. The amount of P and K fertilizers applied decreased from the fourth decade onward.

Table 4.4 Relationship Between Planting and Harvesting Date, Planting Density, and NPK Fertilizer Levels with Dryland and Irrigated Corn Yields

Crop and Plant Management Factors	Intercept	Coefficient for Yield Change for a Unit Changes in Management	SE	R^2
Dryland Yield (Mg ha^{-1})				
Planting date (days from January 1)	14.0	-0.1000	0.010	0.08
Harvesting date (days from January 1)	10.1	-0.0100	0.010	0.02
Density (plant ha^{-1})	0.94	0.0001	0.001	0.36
N level (kg ha^{-1})	3.97	0.0300	0.001	0.43
P_2O_5 level (kg ha^{-1})	5.42	0.0400	0.010	0.12
K_2O level (kg^{-1})	5.97	0.0300	0.010	0.04
Irrigated Yield (Mg ha^{-1})				
Planting date (days from January 1)	19.84	-0.0700	0.010	0.09
Harvesting date (days from January 1)^a				
Density (plant ha^{-1})	2.46	0.0001	0.000	0.44
N level (kg ha^{-1})	8.77	0.0010	0.001	0.11
P_2O_5 level (kg ha^{-1})	10.32	0.0170	0.005	0.03
K_2O level (kg ha^{-1})	10.91	-0.0350	0.016	0.01

^{a}The relationship between irrigated yields and harvesting date was not significant at $P \leq 0.05$.

4.2.2.3 Climatic Factors that Contribute to Yield Changes

When climatic factors were analyzed, a significant positive correlation was found between March and July rainfall and dryland yields (Table 4.5). A significant positive relationship also was found between March minimum and maximum temperatures and May through July minimum temperatures and dryland yields. On the other hand, high maximum temperatures in July through September were negatively correlated with dryland yields.

Although correlation does not necessarily imply cause and effect, it is not difficult to identify potential linkage between these climate factors and dryland corn yield. The positive relationship between higher rainfall and high temperature in March with dryland yields may indicate conditions that favor earlier planting. Early planting date in turn positively correlates with yield, as indicated in Table 4.4. July rainfall coincides with a time in which corn has high demand for water, which explains its

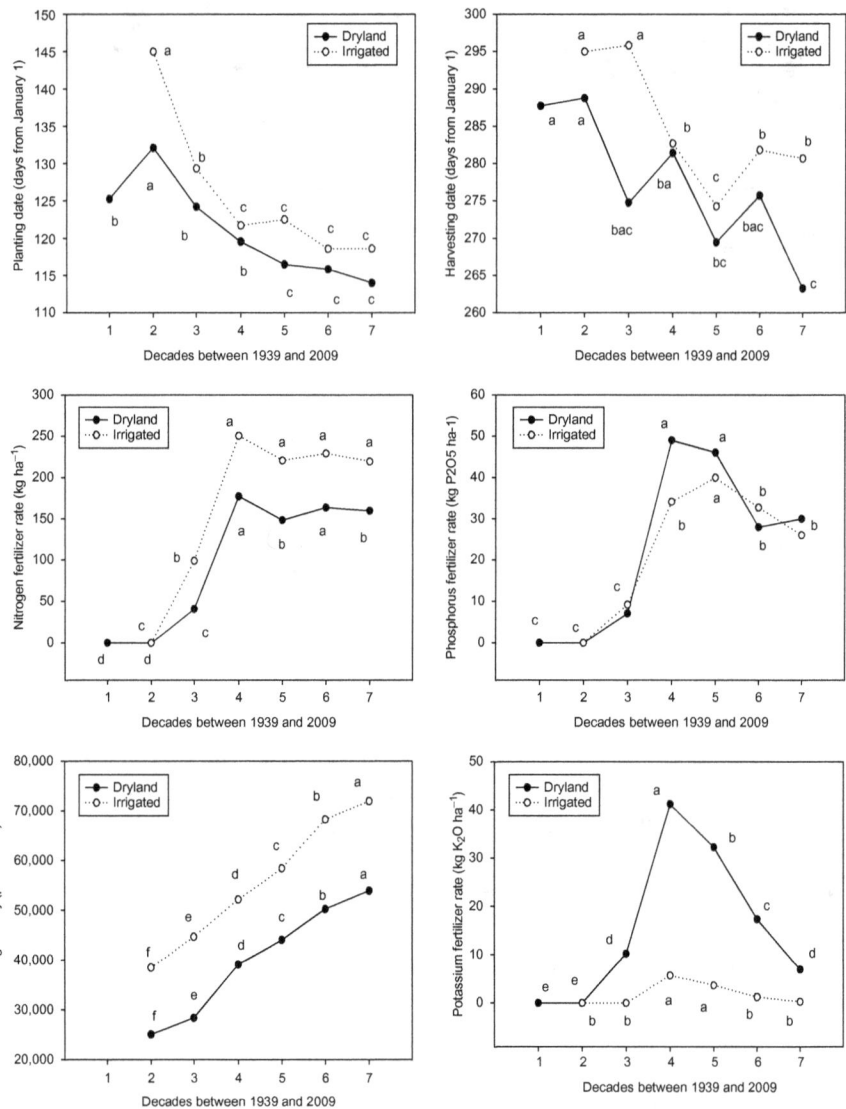

Figure 4.3 Decadal trend in dryland (solid line) and irrigated (dotted line) corn hybrid trial planting and harvesting date, planting density, and NPK fertilizer levels from 1939 to 2009. Similar letters within a line of each management shows a nonsignificant difference between decades.

significant impact on dryland corn yield. High minimum temperature from May through July favors germination and seedling development, and may be associated with a decreased likelihood of a late freeze. On the other hand, high maximum temperatures in the months of July through September may have a negative impact on seed set and grain

Table 4.5 Pearson Correlation Coefficient (R) for Significant Relationship Between Average Yield with Rainfall and Maximum and Minimum Temperatures from 1939 to 2009

Variables	Pearson Correlation Coefficient (R)						
	March	April	May	June	July	August	September
Dryland							
Rainfall and yield	0.26**				0.32***		
T_{max} and yield	0.21*				-0.43***	-0.27**	-0.19*
T_{min} and yield	0.37***		0.34***	0.25**	0.19*		
Irrigated							
Rainfall and yield							
T_{max} and yield	0.28***					-0.17*	
T_{min} and yield	0.18*						

**Significant at $P \leq 0.001$.
**Significant at $P \leq 0.01$.
*Significant at $P \leq 0.05$.

fill (Wilhelm et al., 1999), usually correlates with dry weather at a time when water is most required, and therefore decreases yield.

Correlation analysis revealed fewer significant correlations between climate factors and irrigated corn yields. No significant correlation was found between rainfall and irrigated corn yields. However, March minimum and maximum temperatures were positively correlated with irrigated yield. This is similar to the relationship observed with dryland yields and may again be related to earlier planting. High maximum temperatures in August were correlated with decreased irrigated yields, and possibly indicating high evapotranspiration demand that might decrease the ability of irrigation to meet the demand of the crop in addition to the direct negative effect of high temperature on grain fill (Wilhelm et al., 1999).

The relationship between seasonal weather parameters and yield agrees with previous reports by other authors but also has a unique nature. Tannura et al. (2008) reported a positive impact of high May and July rainfall and a negative impact of above-average July and August temperatures on corn yield. Similarly, Lobell and Asner (2003), Hu and Buyanovsky (2003), and even early studies by Smith (1903) concluded that warmer temperatures early in growing season (April to June) and a wetter May to August interval are favorable for corn yield. However, all these studies looked at either average monthly temperature or daily range of temperature. This study specifically shows which components of seasonal temperature relate to greater yield, i.e., relatively high minimum temperature early in the season (May and June); high minimum temperature, lower maximum temperature, and high rainfall in mid-season (July); and lower maximum temperature (not minimum temperature) late in season (August to September) positively relate to yield.

Trend analysis for total monthly rainfall and average monthly maximum and minimum temperature revealed differences in climate change between regions and months. For the three regions with only 25 years of climatic data (1985–2009), no significant change was detected in any of climatic variables for any of the 12 months. For the six regions with climatic data for the entire hybrid era (1939–2009), trend analysis detected changes for at least one climate parameter in at least one month.

A significant increase in March rainfall was detected for southeast Kansas over the 70 years. There was no trend in rainfall for any month in any of the other regions. A significant positive trend for average monthly minimum temperature was found for the months of March and May through July in northeast Kansas. On the other hand, contrasting trends in average monthly maximum temperature were detected for different months in different regions. An increasing trend in average March maximum temperature was detected for northeast Kansas and a decreasing trend in average monthly maximum temperature was found for September in east-central Kansas, July through October in southeast Kansas, and August through October for northwest Kansas.

These findings on climate change in Kansas agree with previous studies and disagree with others. The Environmental Protection Agency (EPA) (1998) reported a 20% change in precipitation in south eastern Kansas for the time 1900–1998, which agrees with the present study. Our result also agrees with climate change models that predict an increase in precipitation in eastern Kansas (Ojima and Lackett, 2002). However, EPA also reported about a 5–10% increase in precipitation for other parts of Kansas, which differs from our findings. This discrepancy might be due to a difference in length of years of data used.

Evidence shows that global temperature has been rising at a rate of 0.2°C decade^{-1} for the last 30 years (Hansen et al., 2006). The potential positive impact of climate change scenarios on corn yield in the United States, Canada, and China, was recently reported (Li et al., 2011). We did not detect climate change in all regions. Other research reports also suggest that climate change might not be detected in all regions. For example, Tannura et al. (2008) concluded no upward or downward trend in temperature and precipitation at sensitive stages of corn development for Illinois, Indiana, and Iowa. Likewise, no trend was detected for precipitation and temperature from 1895 to 1998 for Missouri (Hu and Buyanovsky, 2003), and Assefa and Staggenborg (2010) reported no significant weather change (temperature and precipitation) in sorghum growing seasons in different areas of Kansas for the past 50 years.

4.2.2.4 Genetic Factors that Contribute to Yield Changes

Besides management and climate, hybrids have changed significantly from 1939 to 2009. Changes in genetics were analyzed in terms of changes in the percentage of new hybrids included in the trial compared with the preceding year and the preceding decade. Analysis of the Pioneer brand hybrids entered in the corn performance trials revealed that out of the average of 10 hybrids entered each year, about 33% were new compared with previous years, and 87% were new compared with the previous decade (Table 4.6).

Table 4.6 Mean Separation Test for Total Number of Pioneer Hybrids Entered in the Corn Hybrid Test, and Percentage of New Hybrids from the Total Entered Compared Each Year and Each Decade

Decade	Year Range	Total Number of Pioneer Hybrids Entered†	New Hybrids for the Year (% from Total Entered)	New Hybrids for the Decade (% from Total Entered)
1	1939–1948	5.3d	20.5b	–
2	1949–1958	9.0bc	22.3b	58.5c
3	1959–1968	10.9ba	29.6b	89.0ba
4	1969–1978	13.8a	37.0b	93.6ba
5	1979–1988	7.6dc	33.4b	86.4b
6	1989–1998	8.9bc	32.7b	97.5ba
7	1999–2009	11.8ba	56.0a	100.0a
	Average	10	33	87

† Within columns, means followed by same letter are not significantly different at $P \leq 0.05$.

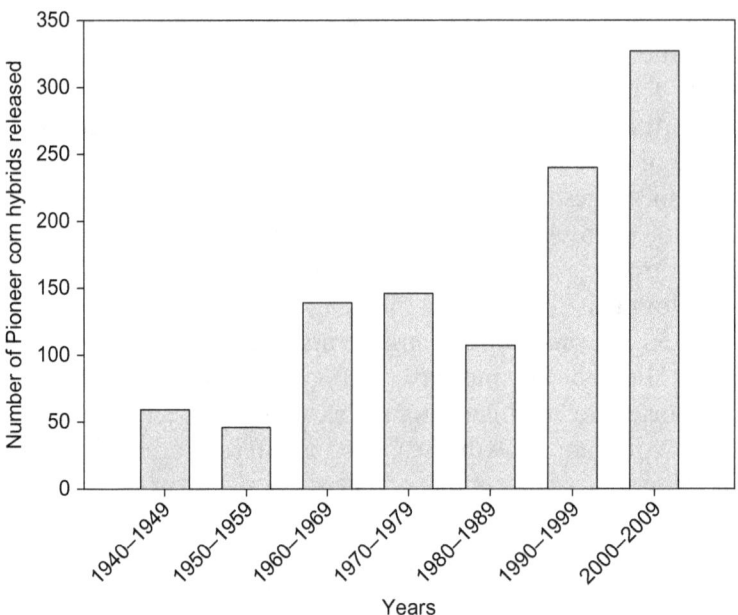

Figure 4.4 Total number of corn hybrids released to North America by Pioneer Hi-Bred, Intl. from 1940 to 2009. Data provided by Keaschall J., research director, Pioneer Hi-Bred Intl., 2011 personal communication.

A decade-by-decade analysis of hybrids revealed that the first decade had the lowest average total number of hybrids entered and had one of the lowest numbers of new hybrids each year within that decade. The total number of new hybrids was also low for decade 2. The number of new hybrids in decade 2 compared with hybrids in decade 1 was only 60%. This means 40% of the hybrids used in decade 2 were the same hybrids that existed in the first decade. The number of new hybrids compared with hybrids in previous decades increased to more than 85% for the remaining decades with complete turnover of hybrids from decade 6 to 7.

In order to substantiate this data about genetic change over the years, the approximate number of hybrids released every five years by Pioneer Hi-Bred, Intl. for North America is plotted in Figure 4.4. The figure suggests that the numbers of hybrids released over the decades have increased substantially, from which we can safely infer high genetic turnover in recent years.

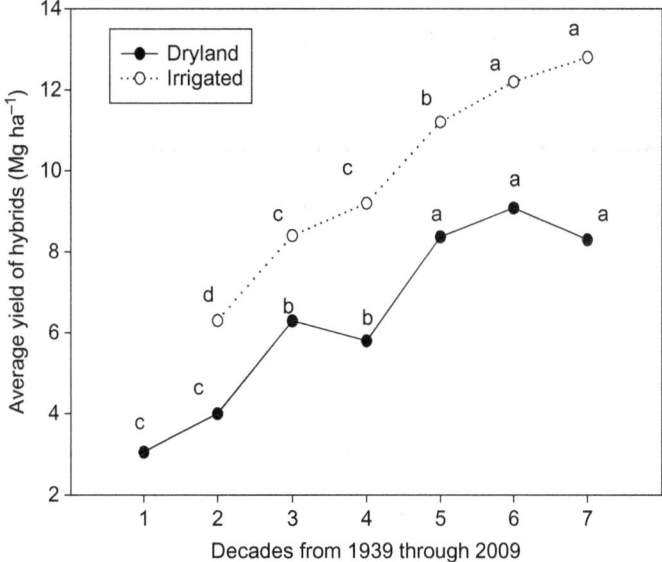

Figure 4.5 Decadal trend in dryland (solid line) and irrigated (dotted line) corn yields in Kansas Grain Performance Trials. Similar letters within a line show a nonsignificant difference between decades.

4.2.2.5 Decadal Changes in Yield

No significant dryland yield change occurred between the first and the second decade (Figure 4.5). This could be due to a number of factors, such as relatively smaller changes in genetics (Table 4.6), no changes in NPK fertilizer, and similar harvest date in decade 1 and decade 2 (Figure 4.3). Changes in planting density, planting date, and about 60% hybrid turnover that occurred between the first and second decade were not associated with significant yield changes.

A significant yield increase occurred in the third decade in both irrigated and dryland corn compared with the first two decades (Figure 4.5). Many factors may be responsible for this change. Significant changes occurred in planting density, planting and harvest date, NK fertilizer (Figure 4.3), and changes in hybrid (Table 4.5) may have contributed to yield increase.

The observed changes in planting density, planting date, NPK (Figure 4.3), and about 93% of hybrid changes (Table 4.5) that occurred between the third and fourth decade were not associated with a concurrent significant increase in dryland or irrigated corn yield (Figure 4.5).

Another significant jump in yield occurred in the fifth decade in both irrigated and dryland corn (Figure 4.5); however, no significant changes occurred in many of the management factors (planting and harvesting date, N) except for planting density, so this change in yield is likely due to genetics and climate (for some regions).

Dryland and irrigated corn trends differed during the past three decades. Dryland corn yield showed no significant change. No significant change was found in N fertilizer, planting date, or harvest dates for that period, but P and K fertilizer rates declined. Planting density, hybrid improvements, and other factors such as climatic changes were potential contributors to yield increases from the fifth decade onward for irrigated corn.

Our results are consistent with previous findings. Duvick (2005) stated that the yield increase in the 1960s and 1970s compared with the previous decades can be explained by the large number of technological changes and hybrid adoptions after World War II. The yield jump in the 1980s compared with the 1960s and 1970s was also attributed mainly to a continued increase in hybrid technologies. The fact that mean yield did not change for the past three decades in dryland may indicate that in the absence of a technological breakthrough, we might be approaching a yield plateau. A reduced yield increase in the late 1970s and early 1980s was evidenced by Menz and Pardey in 1983, but they did not conclude the probability of approaching a yield plateau at that time due to anticipation of more exploitation of genetic diversity and the unpredictable effects of emerging biotechnologies. Similarly, Garcia et al. (1987) acknowledged the declining rate of yield increase and suggested that a plateau might ultimately be reached if no technology breakthrough occurred.

4.3 GRAIN SORGHUM FROM 1957 TO 2009

4.3.1 Data Organization and Analytical Methods for Sorghum

Multilocation grain sorghum hybrid performance trials have been conducted since 1957 in Kansas. We used performance trial data from Greely/Wichita and Finney counties for irrigated yield analysis and data from Ellis, Reno, Finney, and Thomas counties for dryland yield analysis (Table 4.7). These sites were selected because performance trials have been conducted at each of these locations from approximately 1957 to 2008.

Table 4.7 Mean Yield of All Hybrids and the Highest Yielding Hybrid in Grain Sorghum Hybrid Performance Trials Conducted from 1957 to 2008 at Five Locations in Kansas

County	Experimental Station Coordinate	Elevation (m)	Dryland/ Irrigated	Mean Yield of Hybrids (Mg ha^{-1})	Mean Yield of Highest Yielding Hybrid (Mg ha^{-1})	Overall Mean of Hybrids (Mg ha^{-1})	Overall Mean of Highest Yielder (Mg ha^{-1})
Greeley/ Wichita	38°28'13"N, 101°45'16"W	1,101	Irrigated	7.76a†	9.37a†	7.71a	9.33a
Finney	37°58'55"N, 100°51'22"W	872	Irrigated	7.67a	9.29a		
Ellis	38°52'46"N, 99°19'20"W	650	Dryland	5.49b	6.84b	4.81b	5.89b
Reno	38°3'56"N, 97°55'25"W	487	Dryland	4.99b	6.14c		
Finney	37°58'55"N, 100°51'22"W	872	Dryland	4.89b	5.61dc		
Thomas	39°23'32"N, 101°2'51"W	973	Dryland	3.87c	4.99d		

† Within columns, means followed by same letter are not significantly different at $P \leq 0.05$.

Greely and Wichita are 50 km apart in western Kansas and have similar environmental conditions. The irrigated sorghum performance trials were first started at Wichita for 2 years and continued in Greeley. Therefore, in this paper, Greely/Wichita represents the data set from the two counties. The performance trials at Greely, Finney, Reno, Ellis, and Thomas counties were located at the experimental stations at Tribune on Ulysses (Fine-silty, mixed, superactive, mesic Aridic Haplustolls) silt loam soil, Garden City on a Kelith (Fine-silty, mixed, superactive, mesic Aridic Argiustolls) silt loam, Hutchinson on an Ost (Fine-loamy, mixed, superactive, mesic Udic Argiustolls) loam, Hays on a Harney (Fine, smectitic, mesic Typic Argiustolls) silt loam, and Colby on a Kelith silt loam soil, respectively.

Two yield data sets were used from the trial sites: (i) average yield of all hybrids tested, and (ii) yield of the top-performing hybrid from each year. Analysis of yield over years and mean separation using Duncan's multiple tests were conducted by location and for the two data sets using the PROC GLM procedure of the commercial statistical software package SAS (SAS Inst., Cary, NC, 2001).

Average yield was regressed against years for each county from 1957 to 2008 using the PROC REG procedure in SAS. Regression analysis was also conducted for both average yield and the top-performing hybrid yield by combining all dryland counties as one data set.

Agronomic management practices changed over time. Planting date, fertilizer use, and planting density (distance between rows and plants) were recorded for each site-year in performance trial reports (Kansas grain sorghum performance trial reports, 1957–2008). Regression and correlation analyses were conducted for these three agronomic factors for their effect on yield and in relation to years. The analyses were completed using the PROC REG and PROC CORR procedures of SAS. Nitrogen fertilizer application had a significant correlation ($P \leq 0.05$) with both yield and years in dryland trial sites. To single out the contribution of nitrogen fertilizer application to yield increase, analysis of covariance (ANCOVA) was conducted in SAS, i.e., for the combined dryland data; yield was equated to N rate plus hybrid in the model statement, where hybrid was the independent variable and N rate was covariate. The resulting least-square means from this analysis were used to regress for the contribution of increasing nitrogen levels for yield increase.

To determine whether the yield response in dryland sites was influenced by precipitation; prior planting and in season, or temperature changes, the total monthly precipitation data for November, December, January, February, March, and April, and both total monthly precipitation and mean monthly temperature data for May, June, July, August, and September were analyzed using the PROC CORR procedure of SAS. The correlation coefficients were used to determine a significant change in the weather trend and relationship between weather factors and yield.

4.3.2 Results

Grain sorghum yields at the irrigated sites, Greeley/Wichita and Finney counties, were higher than hybrids in the dryland performance trials (Table 4.8). The mean yield of all hybrids was 7.7 Mg ha^{-1} at irrigated sites and 4.8 Mg ha^{-1} at dryland sites. Mean yield of the highest yielding hybrid was 9.3 Mg ha^{-1} at irrigated and 5.9 Mg ha^{-1} at dryland sites.

Table 4.8 Changes in Mean Grain Sorghum Yield from 1957 to 2008 at Five Locations in Kansas

County	DF	Intercept (Mg ha^{-1})	Parameter Estimate (Slope) (Mg ha^{-1})	Standard Error (Slope) (Mg ha^{-1})	t-value for Slope	Pr > t for Slope	R^2
Greeley/Wichita	1	7.2	0.017	0.015	1.11	0.27	0.02
Finney	1	7.9	−0.004.	0.009	−0.46	0.64	0.00
Ellis	1	4.2	0.050	0.015	3.44	<0.01	0.19
Reno	1	3.4	0.058	0.019	3.13	0.01	0.16
Finney	1	2.9	0.033	0.011	3.10	<0.01	0.16
Thomas	1	3.9	0.040	0.015	2.59	0.01	0.11
Overall for Irrigated and Dryland Category							
Irrigated	1	7.5	0.006	0.009	0.69	0.48	0.00
Dryland	1	3.6	0.046	0.008	5.70	<0.00	0.13
DF, Degree of Freedom.							

Regression analysis indicated that average sorghum yield has been nearly constant at Greeley/Wichita and Finney counties over the past 52 years (Table 4.8). This lack of yield increase over time agrees with the finding that irrigated sorghum and wheat yields at Tribune, KS, did not increase between 1974 and 2004 (Stone and Schlegel, 2006).

Yield levels and responses were different for each dryland site. The analysis of dryland data revealed an average yield increase of about 46 kg ha^{-1} yr^{-1} between 1957 and 2008 (Table 4.8). Eghball and Power (1995) reported a similar, 50 kg ha^{-1} yr^{-1}, yield increase in their analysis of 61 years of farmers' sorghum yield data reported by National Agricultural Statistics Service (NASS) (USDA, 1930–1990). The analysis of Eghball and Power (1995) was national in scope, included both dryland and irrigated production together, and includes data from earlier than the 1950s'. The similarity of our results indicate that (i) the yield gains were due to an increase in dryland sorghum production through hybrid improvement, and (ii) yield gains were not only due to change from open-pollinated varieties to hybrids but also due to changes within the hybrid era. USDA-NASS data for Kansas for the same period (1957–2008), however, shows a yield gain ≥ 50 kg ha^{-1} yr^{-1}. The discrepancy between the yield gain reported in this paper and USDA-NASS exists because the USDA-NASS data includes data from years with significant nonhybrid production.

Table 4.9 Mean Planting Date, Inter- and Intra-Row Spacing, N and P Application Rates, and Their Correlation with Year and Yield for Grain Sorghum Hybrid Trials Conducted at Four Dryland and Two Irrigated Locations in Kansas Between 1957 and 2008

	Mean Planting Date (mm/dd)	Distance Between Rows (cm)	Distance Between Plants (cm)	Nitrogen Fertilizer (kg ha^{-1})	Phosphorus Fertilizer (kg ha^{-1})
Dryland					
1957–1966	06/07	93.2	23.4	1.4	0.0
1967–1976	06/03	83.3	26.9	7.5	0.0
1977–1986	06/06	80.0	24.9	28.5	0.7
1987–1996	06/01	76.5	21.1	63.8	10.2
1996–2008	05/24	76.2	17.0	86.1	11.8
Correlation with Year	*	*	*	*	ns
Correlation with Yield	ns	ns	ns	*	ns
Irrigated					
1957–1966	06/04	64	7	67.3	0.0
1967–1976	05/27	68	8	142.7	8.3
1977–1986	05/29	76	8	151.4	24.5
1987–1996	05/30	76	8	131.4	14.8
1996–2008	05/25	76	7	124.0	5.1
Correlation with Year	*	*	ns	*	ns
Correlation with Yield	ns	ns	ns	ns	ns

*Significant at 0.05 probability level.

Sorghum hybrids were first available to farmers in 1957 but accounted for 90% of the planted area by 1960 (Smith and Frederiksen, 2000).

Management practices changed over time (Table 4.9). Mean planting date was 1.5–2 wks earlier in recent years compared with that of early years. Row spacing increased over time. Intra-row spacing decreased for the dryland locations with time but did not change at irrigated locations. Nitrogen application rate increased compared with the earliest trials, but mean N rate for the irrigated trials was less in recent trials compared with trials conducted between 1967 and 1996. At dryland locations, N rate was related to mean yield and increased N rates accounted for about 17 kg ha^{-1} yr^{-1} of sorghum yield increase at the dryland trials (Figure 4.6).

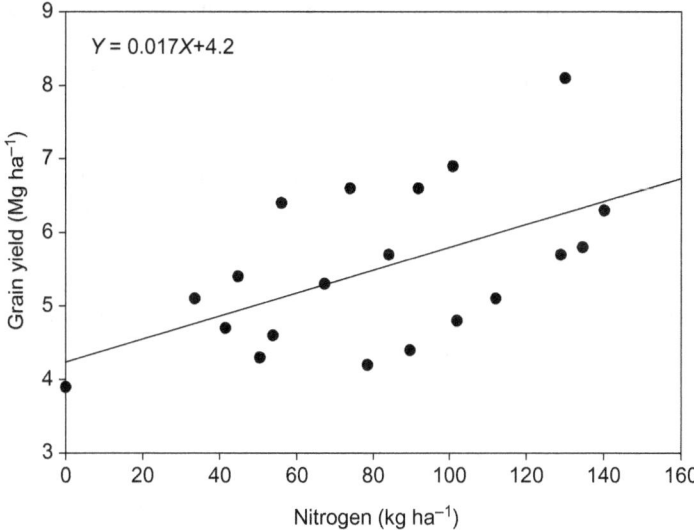

Figure 4.6 Contribution of increase in nitrogen fertilizer to yield in dryland sorghum hybrid performance trials conducted at five locations in Kansas. A regressed least-square means of ANCOVA after adjusting for contribution of other factors in the 52 years analyzed.

Sorghum yield was positively correlated with July and August rainfall and negatively correlated with August temperature perhaps because these months are when sorghum has the greatest water and nutrient demand and when anthesis and grain fill occur. These results agree with those of Staggenborg et al. (2008), who concluded that there was a positive relationship between yield and June to August precipitation and a negative relationship between yield and June to August temperature.

A significant ($P \leq 0.05$) positive correlation between precipitation prior to planting and years was found only for January and April. However, there was no correlation between precipitation prior to planting for the months of November, December, January, February, March, and April and grain yield (data not presented). The Pearson correlation coefficients for the relationship between years and in season monthly precipitation or temperature for May, June, July, and August were not significant ($P \leq 0.05$), but there was a significant overall decrease in September rainfall from 1957 to 2008. The change in September rainfall and temperature, however, was not related to grain yield. Therefore, an increase in precipitation or change in temperature over the time period of our analysis was not considered a factor that contributed to the increasing yield response in the dryland sites over the time period analyzed.

A dramatic increase in atmospheric CO_2 concentration has been reported for the globe since the industrial age (Houghton et al., 1990). Research showed that there is an increase in grain yield of sorghum due to elevated CO_2 under drought conditions (Ottman et al., 2001). However, as sorghum is a C_4 grass, the response to increased CO_2 is smaller than C_3 species (Poorter et al., 1996). Assessing the contribution of increased CO_2 is beyond the scope of this paper, as it is difficult to find CO_2 changes for the locations and years that we considered. However, we acknowledge that there could be a yield gain associated with increased CO_2 confounded with other factors in the drylands.

A number of other factors judged to be yield-limiting were reported in the annual reports including: drought or prolonged dry periods; delayed rainfall; different pests and diseases; cool, wet weather at planting or harvest; lodging; excessive or erratic rainfall; early frost; snow and extreme cold conditions; washing rain and hail; high temperature; hot, dry summer; and high-windy conditions (Kansas Grain Sorghum Performance Trial Reports, 1957–2008). Some pests and diseases arose or become more serious during the years of these trials including greenbug (*Schizaphis graminnum*) which first occurred in 1968 with devastating effects. Development of genetic resistance prevented a loss in hybrid yield potential. Considering these potential yield-limiting factors shows that the genetic yield gain is more than indicated by yield trends reported here. However, we assumed that these new pests and disease occurrences are common to both irrigated and dryland fields; therefore, our comparison is sufficient to compare dryland and irrigated yields.

The gap between irrigated and dryland sorghum yields is narrowing as dryland yields have increased while irrigated yields have remained almost constant. Nitrogen fertilizer explained 34% of the dryland yield increase (Figure 4.6). The remaining yield increase was due to improved hybrids and their interaction with agronomic practices and perhaps with environmental factors like elevated CO_2. The fact that yield has increased in dryland areas but remained constant in irrigated areas shows that hybrid improvement programs were, knowingly or unknowingly, selecting hybrids with better drought tolerance in addition to other characteristics such as insect pest resistance. Testing of prereleased hybrids in more dryland compared to irrigated-sites likely favored selection for enhanced drought tolerance.

The general yield improvement with advancement of sorghum hybrids in the United States that was reported by others (Duvick, 1999; Eghball and Power, 1995) is, therefore, mainly due to enhancement of sorghum hybrids and fertilizer use, which help overcome drought effects.

4.4 CONCLUSION

The hybrid era for corn (1939 to present) started about 20 years earlier than sorghum (1957 to present). Corn hybrid performance trial analysis revealed a positive yield trend for the past 70 years, 1939 through 2009, for both dryland and irrigated corn (Figure 4.5). On an average, the rate of yield increase was about 105, 15 kg ha^{-1} yr^{-1} higher for irrigated and 15 kg ha^{-1} yr^{-1} lower for dryland from this average. On the other hand, our analysis indicated that the average dryland yield increase for grain sorghum was about 46 kg ha^{-1} yr^{-1}, between the years 1957 and 2008, but no significant yield changes were observed in irrigated sorghum yields.

We conclude that sorghum hybrid development has been most effective for dryland sorghum production, possibly because of improved tolerance to water deficits. Further investigation of traits that helped recently developed hybrids better tolerate water deficits indicated that increased root biomass was among qualities selected with advancement of sorghum hybrids (Assefa and Staggenborg, 2011). Improvement of sorghum yield potential for irrigated areas might be enhanced by increased evaluation and selection of hybrids under adequate soil water conditions.

Future attempts to increase yield should build on strategies from the past. First and foremost, irrigable yields were constantly higher than dryland yields, in both crops, suggesting any soil and crop management system that increases available water to crop could contribute to better yield. This includes residue management practices that can help harvest water and snow (Klocke et al., 2009) and notill or reduced tillage operations. Because early planting and harvest has positive yield advantage, due to more reliable rainfall in spring, planting dryland corn and grain sorghum earlier coupled with early cold-tolerance research should be explored further. Continuing intensive hybrid improvement research with the help of advanced biotechnological techniques also should be emphasized. Identifying yield-limiting nutrients at each location and supplying nutrients using site-specific management techniques will help increase yield and cope with future increases in global population.

CHAPTER 5

Yield Distribution and Functions for Corn and Grain Sorghum

Grain yield is among the top factors that influence producers' selection of crops. A number of statistical models have been developed to describe crop yield in relation to environmental and technological factors. The majority of statistical crop models developed before the late 1950s relate crop yield with climatic conditions alone (Compton, 1943; Mathews and Brown, 1938; Sanderson, 1954). Models released after the 1950s started to incorporate the influence of technological and management factors in yield models (Shaw, 1964; Oury, 1965; Kaylen and Koroma, 1991; Nelson and Dale, 1978; Thompson, 1975; Schlenker et al., 2004; Lobell and Asner, 2003).

Correlation and regression analysis of corn yield with rainfall, seasonal temperature, and management factors have been reported by different authors (Assefa et al., 2012; Runge and Odell 1958; Schlenker and Roberts, 2009; Tannura et al., 2008; Thompson, 1969). Few correlation and regression models are available for grain sorghum response to various weather and management factors (Assefa and Staggenborg, 2010; Seiler, 1984; Hodgest et al., 1979). Our objectives in the present study were (i) to better understand corn and grain sorghum yield distribution, (ii) to partition the variability of yield explained by genetics and environment, and (iii) to model and compare yield responses of these two crops.

5.1 DATA ASSEMBLY AND ANALYTICAL STEPS

The data for this research were assembled from Kansas Corn Performance Trials and Kansas Grain Sorghum Performance Trials conducted in 11 counties of Kansas from 1992 to 2012 (Figure 5.1). In these trials, corn and sorghum hybrids were planted every year and were evaluated for yield and other traits. Roughly, half the entries in any test in any year consisted of hybrids tested in previous years, with the balance consisting of new hybrids (Lingenfelser et al., 2012). A subset of three maturity check hybrids was consistent across all locations in each year. These check hybrids changed every few years to reflect the latest genetics, but only one hybrid was replaced in any one year. This entry structure provided a level of year-to-year and location-to-location consistency balanced with an annual infusion of new entries that reflected the latest genetic developments. In addition to yield data, hybrid names, the amount of nitrogen (N) fertilizer applied, planting and harvesting dates, cropping system (irrigated or dryland), average daily maximum and minimum temperatures, and average daily rainfall data were assembled.

Datasets were assembled for corn and sorghum separately in long data format to include name of each county (space), years the data were collected (time), names of the hybrids, yields of each hybrid, amount of N fertilizer applied, length of growing season, cropping system (irrigated or dryland), rainfall, and minimum and maximum temperatures for the months from April to September. County and year were used to create a classification variable designated as an environment that captured both space and time components.

Analysis was performed in R (R 2.15 statistical program) and, when necessary, in SAS. First, the distribution of irrigated and dryland yield for each crop was determined by plotting yield values in the x-axis and their frequency in the y-axis. The mean and standard deviation (SD) of yield for dryland and irrigated corn and sorghum were also calculated.

The variability in yield was explained by genetics and environment and their interaction was partitioned using a MIXED model with random environment and hybrid variables. Initial analysis revealed that

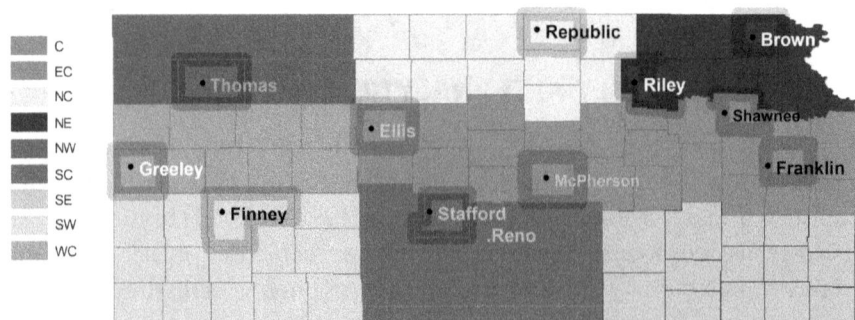

Figure 5.1 Map of Kansas depicting counties and their districts from which the Kansas Corn and Sorghum Hybrid Performance Trials data for the years 1992–2009 were assembled.

environment explained the largest variation in crop yield. In the next step of the analysis, environment was split into space and time. Within space and time, environmental and management factors that might directly affect crop yield were listed based on prior knowledge. Yield was modeled against important weather and management factors using multiple linear regression and robust locally weighted regression and smoothing techniques. The significance of incorporating time and space information as explanatory variables to improve model fit was demonstrated.

5.2 GENERAL YIELD DISTRIBUTION

Overall (dryland and irrigated combined) distributions of corn and grain sorghum yields across Kansas are presented in Figure 5.2. Yields of both crops exhibited approximately normal distributions, with both the mean yield and the SD being greater for corn than for grain sorghum. Because both the mean and SDs were different for each crop, the coefficient of variation (CV) was calculated to compare the two crops. Results suggested that overall grain sorghum yields had slightly less variability than corn.

To compare the yields of the two crops by cropping systems, yield distributions of irrigated and dryland corn and sorghum yields were also analyzed (Figure 5.3). Both the irrigated and dryland yield distributions of these crops exhibited an approximately normal distribution, with greater mean and SD in both dryland and irrigated corn compared with dryland and irrigated grain sorghum. The CV was greater

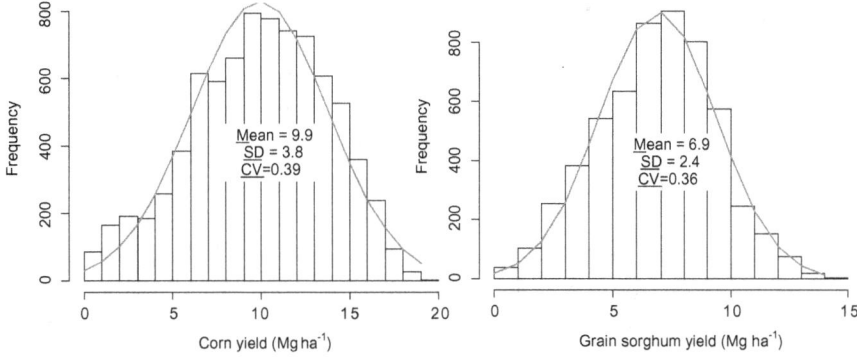

Figure 5.2 Distributions of corn and grain sorghum yields from Kansas Hybrid Performance Trial data.

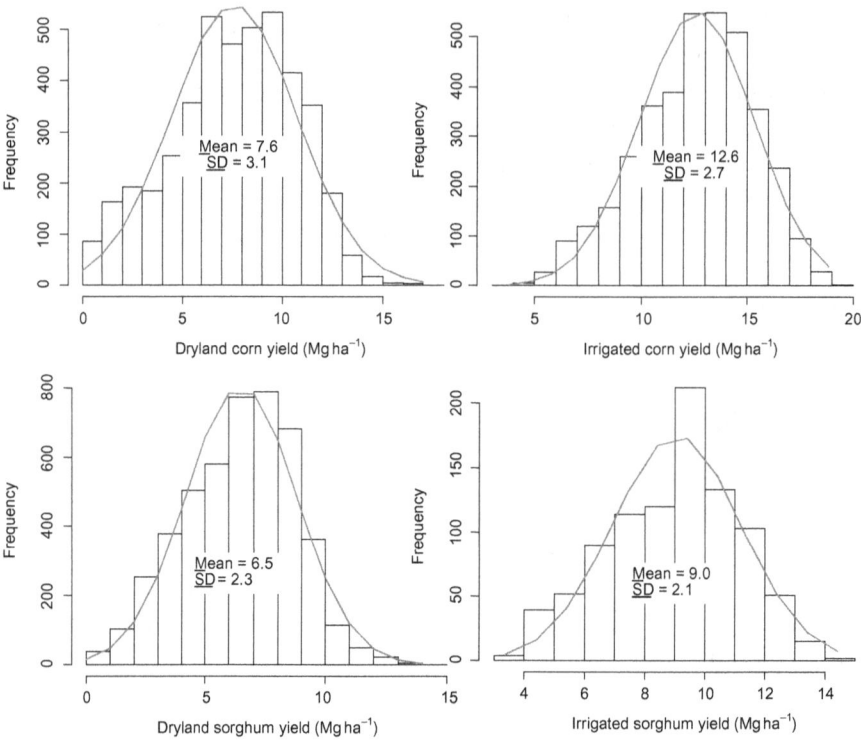

Figure 5.3 Distributions of dryland and irrigated corn and grain sorghum yields from Kansas Hybrid Performance Trial data.

for dryland corn (CV = 0.41) than for dryland sorghum (CV = 0.38), but it was less for irrigated corn (CV = 0.21) than for irrigated sorghum (CV = 0.23).

5.3 PARTITIONING SOURCES OF VARIABILITY

How much of the variability in corn and sorghum yields can be explained by environment or genetics is an important question. Given that yield data for each cropping system (dryland or irrigated corn or sorghum) represent multiple hybrids at each location and year, therefore, hybrid (genetics) is one source of variation. Yield data for each year included multiple locations, making space for another source of variation. The data included 20 years, so time is a third source of variation. To simplify the initial analysis, space and time were combined and designated as environments, which reduced the sources of variation to two: environment (space and time) and hybrid (genetics).

Yield Distribution and Functions for Corn and Grain Sorghum 61

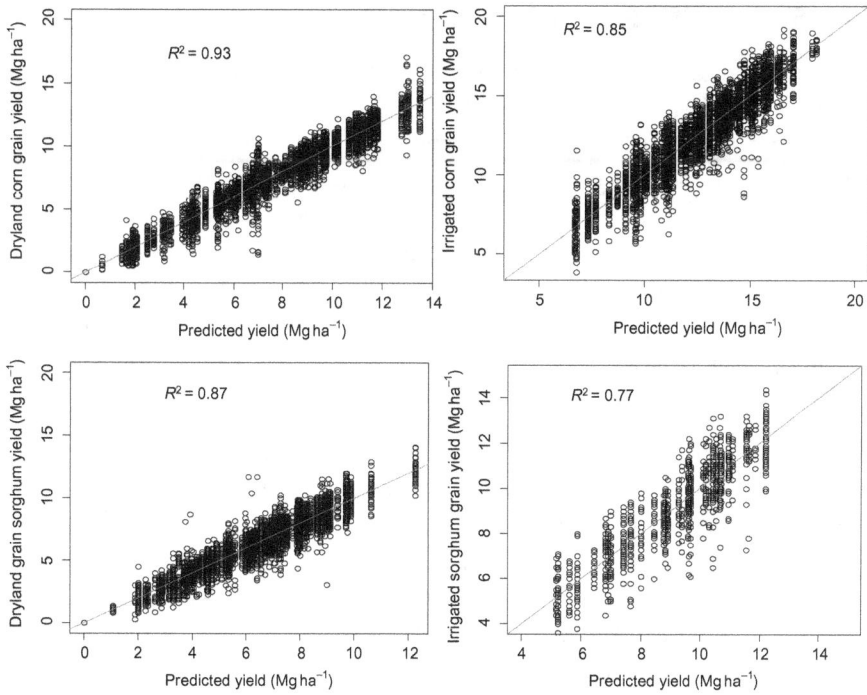

Figure 5.4 The relationship between observed and fitted yield values of a model with fixed environment and random hybrid effects (environment-specific model).

How much of the variability was due to genetics or environment can be estimated by fitting a mixed model of crop yield as a function of fixed environment and random hybrid effects

$$Y_{ij} = \mu_i + H_j + \varepsilon_{ij} \tag{5.1}$$

Y_{ij} is the yield at environment i from hybrid j; μ_i is the mean for the ith environment; H_j is the differential random effect of jth hybrid; and ε_{ij} is the residual or interaction effect of jth hybrid and ith environment.

Figure 5.4 presents the relationship between observed yields and fitted values of the model. This environment-specific model explained the variability in yield with coefficient of determination (R-square) values of 0.93, 0.87, 0.85, and 0.77 for dryland corn and sorghum and irrigated corn and sorghum, respectively. This is equivalent to saying environment alone explains 93%, 87%, 85%, and 77% of the variability in yield in dryland corn, sorghum, irrigated corn, and sorghum, respectively.

5.4 ENVIRONMENT COMPARISON

Environments from which dryland corn (105 environments) and dryland sorghum (126 environments) data were collected were separated into four categories within each crop. Categories were high-, medium-, low-, and very low-yielding based on mean yield records of >7.5, 5–7.5, 2.5–5, and <2.5 Mg ha^{-1}, respectively. These environments were compared for weather and management differences (Table 5.1). No consistent significant management differences (NPK fertilizer or length of growing season) were observed among high- to very low-yielding environments, but significant differences were observed in terms of climatic variables (Table 5.1).

The weather in which low to very low (0–5 Mg ha^{-1}) dryland corn yields were recorded was characterized by high maximum and low minimum mean temperature in July (>34, <9.1°C) and August (>32.5, <11°C) and low total monthly rainfall for the months of June (<67), July (<77), August (<61), and September (<47) in millimeters (Table 5.1). The weather for very low (0–2.5 Mg ha^{-1}) dryland sorghum yield was a consecutive three-month high temperature in June (>32), July (>35), and August (>32°C), with low total monthly

Table 5.1 Weather Responsible for Different Levels of Dryland Corn and Sorghum Yields

Dryland Corn Yield (Mg ha^{-1})	Maximum Temperature (°C)		Minimum Temperature (°C)		Total Monthly Rainfall (mm)			
	July	August	July	August	June	July	August	September
0–2.5	35a[†]	33.7a	9.1b	11b	63.3b	49.3b	51.5	23.8b
2.5–5	34ab	32.6ab	9.3b	11.7ab	67.5b	77.5ab	61	47.8ab
5–7.5	33bc	32.3ab	10.5ab	12.1ab	111ab	92.5a	78	71.8a
>7.5	32c	31.5b	11.3a	12.7a	132.5a	114a	99.25	63.3a
Dryland Sorghum Yield (Mg ha^{-1})	Maximum Temperature (°C)			Total Monthly Rainfall (mm)				
	June	July	August	July	August	September		
0–2.5	32.2a	35.8a	32.3a	47.5b	52.5	17.5b		
2.5–5	29.9b	33.7b	32.4ab	67.5b	65	47.5b		
5–7.5	30b	33.1b	32.1ab	97.5ab	87.5	77.5a		
>7.5	29.8b	32.0c	31.3b	125a	90	60ab		

[†] Within columns, means followed by the same letters are not significantly different.

rainfall in July (<47), August (<52), and September (<18). A June mean maximum temperature in the 30's and July and August mean maximum temperatures >32 (high) characterized low- to medium-yielding environments for sorghum. High-yielding environments for the two crops were similar with mean maximum and minimum temperatures of June, July, and August <33, and >10, respectively, and high total monthly rainfall in June and July (>90), August (>78), and September (>71 mm). These results suggest that environmental conditions that drive dryland sorghum to a very low (0–40 bushel acre^{-1}) yield are harsher (three months of low rainfall) than environmental conditions that drive dryland corn (two months of low rainfall). No consistent differences in weather conditions were found between low- and high-yielding environments for irrigated sorghum and corn.

5.5 CORN AND GRAIN SORGHUM YIELD MODELS

5.5.1 Yield as a Function of Seasonal and Monthly Rainfall

Crop yield, particularly dryland yield, is highly dependent on the amount of rainfall in the growing season. The relationships between yields of corn and sorghum with total rainfall from April to September are presented in Figure 5.5. These models were developed using a robust local weighted scatter-smoothing regression technique to fit the data. Both corn and sorghum yields demonstrate a nonlinear relationship with total rainfall. Dryland crop yield increased in both crops with different slopes, from approximately 5 to 27 in. of rainfall and stabilized or decreased afterward. Corn yield was more responsive than

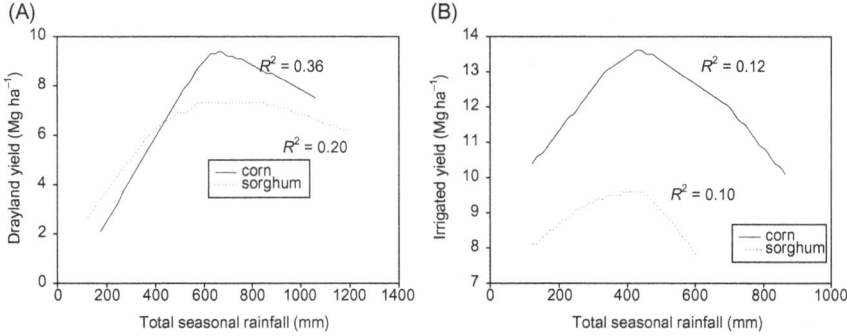

Figure 5.5 Corn and sorghum yields: (A) dryland and (B) irrigated, as a function of total seasonal (April through September) rainfall.

sorghum yield to different levels of rainfall, and dryland yield was more responsive than irrigated yield for both crops. The difference in range of rainfall for irrigated corn and sorghum (Figure 5.5B) suggests that producers do not irrigate sorghum in areas where rainfall is above 600 mm.

Total rainfall alone is not the best variable to model yield because a crop may receive little rain early in the season and get plenty of water after that, but the damage caused by the early rainfall deficit might be irreversible. Therefore, modeling yield using average monthly rainfall may better describe the relationship between rain and yield than total seasonal rainfall.

Table 5.2 presents models for dryland and irrigated corn and sorghum yields using the daily average monthly rainfall and interactions of two consecutive months. Both dryland corn and sorghum yields demonstrate significant positive relationships with daily average monthly rainfall decreasing, in order, in June or July, May, August, April, and September. Dryland yields for both crops are penalized more when two consecutive months have greater rainfall, except for

Table 5.2 Regression Coefficients of Average Daily Rainfall in the Months from April to September and Interactions of Two Consecutive Months in Dryland and Irrigated Yield Models for Corn and Sorghum

Rainfall	Coefficient, Dryland		Coefficient, Irrigated	
	Corn	Sorghum	Corn	Sorghum
Intercept	−1.71***	0.37*	9.84***	6.30***
April	0.22***	0.33***	0.30***	−0.18
May	0.67***	0.55***	0.48***	−0.52***
June	1.30***	0.74***	0.64***	−0.27**
July	1.29***	0.97***	1.11***	2.17***
August	0.53***	0.54***	−0.05	1.48***
September	0.15**	0.02	0.18*	−0.15
April × May	−0.06***	−0.02**	−0.14***	0.12*
May × June	−0.10***	−0.13***	−0.13***	0.07*
June × July	−0.19***	−0.11***	−0.14***	−0.04
July × August	0.05**	−0.06***	−0.06	−0.66***
August × September	−0.09***	−0.02	−0.10***	−0.01
Adjusted R^2	0.43	0.31	0.19	0.25

***, **, * indicate the parameter is significant at 0.001, 0.01, and 0.05 probability levels.

dryland corn for the months of July and August. The greater magnitude of the absolute values of most of these coefficients for dryland corn than for dryland sorghum demonstrates that dryland corn is more responsive to rainfall.

5.5.2 Yield as a Function of Seasonal Temperature

Seasonal temperature is also an important factor in determining crop yield. Crops are grouped into summer and winter crops based on their seasonal temperature requirements. One reason why corn is planted in early April and sorghum is planted in late May is that the crops require different minimum temperatures for germination and growth. Table 5.3 presents estimated coefficients of a linear model that relates yield with the effects of average temperature of months within a season. The coefficients of the model imply that (i) the relationship between average monthly temperature and yield is positive early in the season (April, May) but is negative from June or July to the remainder of the season, (ii) high temperature in July, which has a larger coefficient in both dryland and sorghum models, has a more negative impact on corn than on sorghum, and (iii) average monthly temperature explains only a small portion of yield variability in both crops but is more informative for corn than for sorghum.

The relationships between dryland corn and sorghum yields and temperature can be improved by examining monthly maximum

Table 5.3 Regression Coefficients of Average Monthly Temperature in the Months from April to September in Dryland and Irrigated Yield Models for Corn and Sorghum

Average Temperature	Coefficient, Dryland		Coefficient, Irrigated	
	Corn	Sorghum	Corn	Sorghum
Intercept	20.3***	20.9***	35.0***	18.9***
April	0.16***	0.27***	0.34***	0.19***
May	0.53***	0.07***	0.01	−0.21***
June	0.13*	−0.07	−0.74***	−0.09
July	−0.87***	−0.53***	−0.11**	−006
August	−0.07	−0.03	−0.19***	−0.10
September	−0.13***	−0.13***	−0.08**	−0.11*
Adjusted R^2	0.24	0.11	0.18	0.05
***, **, * indicate the parameter is significant at 0.001, 0.01, and 0.05 probability levels.				

temperatures rather than the average. Maximum temperature might predict dryland yields better than average temperatures due to how an average value dilutes the extremes. Minimum and maximum temperatures are likely to matter more than the average for growth and development. Similar to the average temperature model, (i) relationships between maximum temperature and dryland yields were positive early in the season (April, May) but negative or weakly positive later in the season, (ii) impact of high temperature in July, which has a larger coefficient in both dryland corn and sorghum models, has more negative impact on corn than on sorghum yield, and (iii) the model explains slightly more yield variability than the average monthly temperature model, except for irrigated corn where it explains less (Table 5.4). Even so, the portion of explained yield variability is relatively small for both crops in both dryland and irrigated systems.

An alternative temperature model was developed using maximum temperature with two-way interactions of consecutive months (Table 5.5). Unlike the two previous temperature models, the interaction model suggests a positive correlation between maximum temperature and dryland yield for all months except April and September; however, the model penalizes yield highly as the maximum temperature between two consecutive mid-season months increases.

Table 5.4 Regression Coefficients of Average Monthly Maximum Temperature in the Months from April to September in Dryland and Irrigated Yield Models for Corn and Sorghum

Maximum Temperature	Coefficient, Dryland		Coefficient, Irrigated	
	Corn	Sorghum	Corn	Sorghum
Intercept	37.0***	30.8***	24.7***	14.4***
April	0.18***	0.17***	0.16***	0.02
May	0.32***	0.02*	0.13*	−0.17**
June	−0.17***	−0.17***	−0.59***	0.02
July	−0.76***	−0.46***	0.11***	0.04
August	−0.15***	−0.4***	−0.18***	−0.11*
September	−0.23***	−0.13***	−0.06***	0.00
Adjusted R^2	0.34	0.22	0.11	0.01

***, **, * indicate the parameter is significant at 0.001, 0.01, and 0.05 probability levels.

Table 5.5 Regression Coefficients of Total Monthly Maximum Temperature in the Months from April to September and Interaction of Two Consecutive Months in Dryland and Irrigated Yield Models

Maximum Temperature	Coefficient, Dryland		Coefficient, Irrigated	
	Corn	Sorghum	Corn	Sorghum
Intercept	−174.8***	−119.7***	144.9***	23.2
April	−2.75***	−1.63***	−1.7***	−3.07***
May	3.99***	2.70***	−2.77***	−2.28**
June	5.86***	5.06***	−6.02***	−5.35***
July	3.49***	4.01***	−4.26***	−1.73***
August	2.57***	0.60	2.88***	6.80***
September	−0.89**	−2.25***	3.47***	3.78***
April × May	0.12***	0.07***	0.07***	0.12***
May × June	−0.20***	−0.14***	0.03*	−0.02
June × July	−0.03*	−0.06***	0.14***	0.17***
July × August	−0.10***	−0.08***	−0.00	−0.12***
August × September	0.02*	0.07***	−0.12***	−0.12***
Adjusted R^2	0.40	0.32	0.18	0.10

***, **, * indicate the parameter is significant at 0.001, 0.01, and 0.05 probability levels.

5.5.3 Yield as a Function of Rainfall, Temperature, and Selected Management Factors

Yield functions were further improved by assuming that the relationship between yield and environmental factors is linear with the combination of environmental factors. A yield model was constructed using rainfall, maximum temperature, N fertilizer application, and length of growing season. Figure 5.6 depicts the model fit for the selected model, and Table 5.6 presents parameters and associated coefficients. This model was able to explain about 70% of the variability out of 93% possible variability explained by environment for dryland corn (Figure 5.4), 57% out of 87% for dryland sorghum, 57% out of 85% for irrigated corn, and 56% out of 77% possible variation that could be explained by environment for irrigated sorghum.

The following differences were noted in the dryland corn and dryland sorghum models: (i) June and July rainfall have significant importance for dryland corn, but July and August rainfall were most important for dryland sorghum yield, (ii) warmer and dryer April conditions for dryland corn but cooler and wetter April conditions for

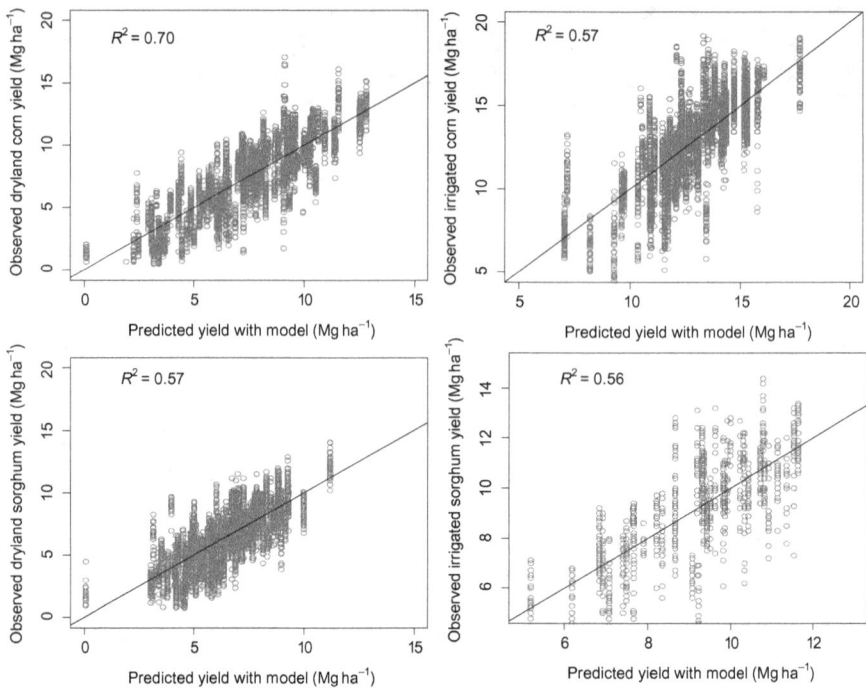

Figure 5.6 Observed yield plotted against fitted values of models that linearly predict yields from total rainfall, average maximum temperature, N fertilizer, and length of growing season as indicated in Table 5.6.

dryland sorghum correlated positively with yield, (iii) cooler, wetter June conditions correlated positively with corn yield, but warmer and slightly wetter conditions in June correlated positively with sorghum yield, (iv) warmer and wetter August conditions were beneficial for corn, but cooler and wetter August conditions were more beneficial for dryland sorghum. These results suggest that the monthly rainfall and temperature conditions that strongly relate to corn and sorghum yields occurred in different months of the season. Therefore, even though the two crops often are considered as competing for summer cropping area, they take advantage of different parts of the season and thus should be considered alternative crops to utilize different environmental conditions.

Analysis of the frequency distribution of yield showed that dryland and irrigated corn and sorghum yields are approximately normally distributed. Mean and maximum possible yield of corn were greater than for grain sorghum. The variation in dryland yield was less for sorghum than for corn.

Table 5.6 Regression Coefficients for Parameters in the Dryland and Irrigated Yield Model

Parameters	Dryland		Irrigated	
	Corn	Sorghum	Corn	Sorghum
Intercept[a]	−53.9	−64.8***	65.5***	89.8
April RF	−0.12***	0.27***	−0.08	−0.55**
May RF	0.27***	0.31***	−0.36***	−0.33**
June RF	0.69***	0.11***	−0.18*	−0.26*
July RF	0.82***	0.72***	−0.27*	1.52***
August RF	0.41***	0.65***	0.25**	1.23***
September RF	−0.04	−0.19***	0.32***	−1.25***
April max	2.21***	−1.26***	−2.52***	4.10***
May max	1.54***	2.71***	−1.47***	−1.82*
June max	−2.24***	4.39***	−0.78	−6.71***
July max	3.03***	2.85***	−1.85***	−5.65
August max	2.81***	−1.22***	0.37	1.64*
September max	−2.60***	−3.17***	1.70***	−2.57
April RF × May RF	0.01	−0.04***	0.09***	0.21***
May RF × June RF	−0.06***	−0.05***	−0.02	−0.11**
June RF × July RF	−0.14***	−0.03***	0.08***	0.11**
July RF × August RF	0.04***	−0.13***	−0.01	−0.82***
August RF × September RF	−0.01	0.01	−0.11***	0.43***
April max × May max	−0.09***	0.05***	0.11***	−0.17***
May max × June max	0.01	−0.13***	−0.02	0.17***
June max × July max	0.04***	−0.05***	0.02	0.06*
July max × August max	−0.14***	−0.05***	0.04*	−0.05
August max × September max	0.07***	0.10***	−0.04***	−0.00
LGS	0.02***	0.02***	0.03***	−0.04***
N	0.01***	−0.01**	−0.01***	0.01***
District Adjustment				
Northeast	−0.38	0.72***	−2.41***	NA
East-central	1.08**	0.31***	1.19***	NA
North-central	−1.01	1.5	3.03***	NA
Central	−0.30*	−0.44*	1.67***	3.1*
South-central	0.87***	−0.68	0.99***	1.83***
Northwest	−0.89	−0.63***	4.25***	2.12***
West-central	0.37	−1.12	2.56***	1.13
Southwest	−0.24***	−0.77	0	0

***, **, * indicate the parameter is significant at 0.001, 0.01, and 0.05 probability levels.
[a]Parameters: average daily rainfall of months from April (APR RF) to September (SEP RF) and their interaction, average maximum temperature of the same months and their interaction, length of growing season (LGS), and applied nitrogen (N).

The variability in yield explained by environment is much greater than the variability explained by genetics within each cropping system, which demonstrates that environment plays a dominant role in determining crop yield. Corn yield variation explained by environment was greater than sorghum, implying greater sensitivity to environmental conditions. The relative stability of sorghum yields across environmental variations compared with corn yields in this analysis supports previous findings (Boyer 1970; Beadle et al., 1973; Stone et al., 1996; Fischer et al., 1982). Obviously, dryland yields are relatively more environmentally dependent than irrigated yields because one environmental factor, water, is less of a limitation in irrigated systems.

High-yielding environments for the two crops were similar, with mean maximum and minimum temperatures of <90°F and >50°F, respectively, in June, July, and August, and high total monthly rainfall in June and July (>5 in.), August (>3.5 in.), and September (>2.5 in.). The environment that drives dryland sorghum to a very low (0–40 bushel acre^{-1}) yield was relatively harsher (3 months of drought) than the environment that drives dryland corn to very low yield (2 months of drought). No consistent weather differences were found for low- to high-yielding environments for irrigated sorghum or corn.

We conducted an exploratory analysis of different yield functions. The initial model was environment specific. Environment was then separated into time and space components. Relationships between yield and different environmental factors were explored, starting with rainfall and adding other environmental factors into the model. In all cases, we demonstrated that corn compared with sorghum and dryland compared with irrigated yields were more responsive to environmental variations.

In the model that combined rainfall, temperature, management, and space adjustments, the monthly rainfall and temperature parameters that strongly related to corn and sorghum yields appeared at different months of the season. Therefore, we concluded that because corn and sorghum take advantage of different parts of the season, they should not be considered to be in competition for summer cropping areas, rather alternative to one another according to environmental conditions.

CHAPTER 6

Resource (Land, Water, Nutrient, and Pesticide) Use and Efficiency of Corn and Sorghum

Crop resource use efficiency can be defined as the ratio of crop output to resource input (Sadras et al., 2007; Blum, 2009; Sinclair et al., 1984; Howell, 2001). Resource use efficiency is an ideal criterion for crop selection for various reasons. First and foremost, resources are scarce and require responsible management. By selecting a crop that performs relatively well with lower inputs of a certain resource, we can extend the life of that resource. Second, almost all resource inputs have an associated cost, and one way to reduce the input cost is to select more efficient crops. Third, some inputs have negative environmental impacts if utilized in an inefficient manner; thus, a crop that requires less of these resources without negatively affecting yield is more environmentally sound.

Land, water, fertilizer, and pesticides are among the most critical resources allocated for crop production. As described above, these resources are scarce and expensive, and some (fertilizer and pesticides) may have a negative impact on the environment if applied in excess (FAO, 1999; Lambin and Meyfroidt, 2011; Rosegrant et al., 2002; Morrison et al., 2009; Pimentel et al., 2005; USGS, 2001). Therefore, knowledge regarding the land, water, fertilizer, and pesticide use efficiency of crops contributes to sound economic and environmental crop production decisions.

Corn and grain sorghum are both warm-season crops adapted for summer-season cropping in the United States. The decision by a producer to cultivate one of these two crops often comes at the expense of not cultivating the other one, so providing objective information to justify decision making for corn or sorghum cropping is crucial. The objective of the present review was to compare corn and sorghum based on their land, water, nutrient, and pesticide use efficiencies. To meet this objective, data from hybrid performance tests, herbicide performance trials, long-term fertilizer trials, journal articles, and extension publications were assembled and analyzed.

6.1 LAND USE EFFICIENCY (YIELD)

Among resources allocated to crop production, cropland is at the top of the list. Yield of a crop, expressed as the weight of grain per area of land, is a measure of crop land use efficiency, which is why crop yield historically is one of the major parameters for crop selection and cropping system decision making. However, among other reasons, if two crops have different production costs, different impacts on the environment, and differ in ease of production, yield may not be the only criterion for crop selection. In this section, we will compare corn and sorghum based on their land use efficiency and formulate a cutoff value assuming similar production costs, impacts on the environment, and ease of production for both crops.

6.1.1 General Yield Relationship

The relationship between sorghum and corn yield in Kansas is depicted in Figure 6.1. This relationship was developed based on data from sorghum and corn hybrid performance trials conducted in close proximity at a given location in a given year from 1992 to 2012. Four different approaches were used to assemble the data and compare the sorghum−corn yield relationship: (i) average yield of all hybrids in dryland trials, (ii) average yield of the five top-yielding hybrids in dryland trials, (iii) average yield of the five top-yielding hybrids in irrigated trials, and (iv) average yield of the five top-yielding hybrids in both irrigated and dryland trials. Using yields from the five top-yielding for each crop eliminates hybrids that may not be completely characterized and may not be adapted to a given location. This approach also captured the greatest yield potential for each crop in a given environment. The different approaches revealed slight differences in the sorghum−corn yield relationship but did not indicate fundamental differences in the nature of the relationship for each situation.

In the relationship between corn and sorghum yields based on average dryland yields of all hybrids in the performance tests (Figure 6.1A), 6 Mg ha^{-1} emerges as a cutoff point. This result implies that in areas where the historical average yield of dryland corn is less than 6 Mg ha^{-1}, sorghum has better average land use efficiency. In areas where historic average dryland corn yield is greater than 6 Mg ha^{-1}, corn has better average land use efficiency. When yields are close to 6 Mg ha^{-1}, both will have similar land use efficiency.

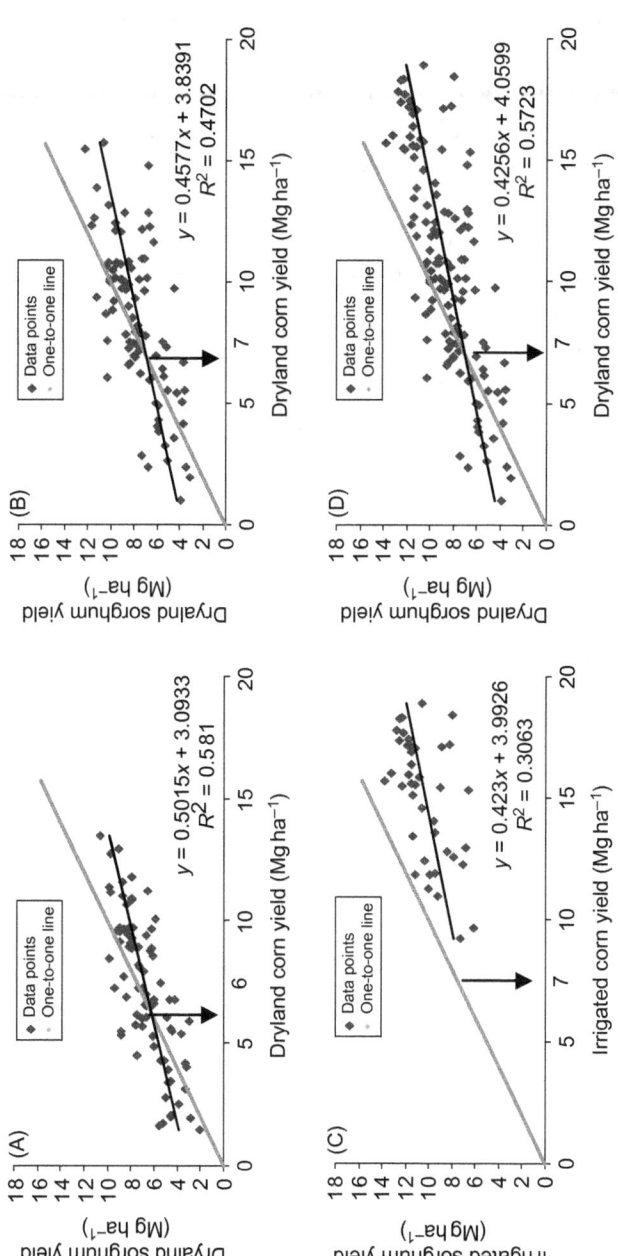

Figure 6.1 Relationship between sorghum and corn yields from Hybrid Performance Trial Data in Kansas from 1992 to 2012: (A) average of all hybrids in dryland trials, (B) average of the five top-yielding hybrids in dryland trials, (C) average of the five top-yielding hybrids in irrigated trials, and (D) five top-yielding hybrids in both dryland and irrigated trials.

6.1.2 Land Use Efficiency Over Time and New Drought-Tolerant Hybrids

A 5-year moving average approach was used to detect possible changes in the relationship between corn and sorghum yields over time. The cutoff point (corn yield value below which sorghum and above which corn has better land use) was calculated for every 5 years from 1992 to 2012 (Figure 6.2). The cutoff value decreased from 8 Mg ha^{-1} for 1992–1996 to 6 Mg ha^{-1} for the most recent 5 years (2007–2012). Fitting a linear model to the 5-year cutoff values revealed a decline of 0.1 Mg ha^{-1} yr^{-1}. This change can be explained by greater increases in corn yields relative to increases in sorghum yields during this time period (Assefa and Staggenborg, 2010; Assefa et al., 2012). This greater yield increase for corn in these environments might be an indicator that recently released corn hybrids have greater tolerance to dryland production conditions than those released in the past; however, this might be due to changes in production practices (e.g., high-residue notill systems).

Release of new drought-tolerant (DT) hybrids such as Optimum AQUAmax, DroughtGard, and Agrisure Artesian by DuPont-Pioneer, Monsanto, and Syngenta, respectively, are expected to significantly improve the drought tolerance of corn. Because these technologies are

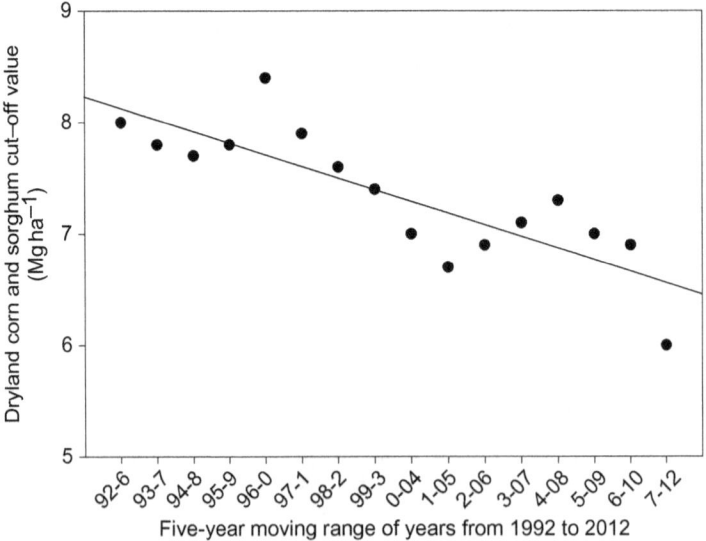

Figure 6.2 Dryland corn and sorghum cutoff point in a 5-year moving average for the years 1992–2012.

new, limited data are available to compare them with existing sorghum hybrids. The relationship between regular dryland corn hybrids and new DT hybrids from limited data available from experiments at Etter, TX, and Topeka, KS, in 2011 and 2012 is depicted in Figure 6.3.

As can be seen in Figure 6.3, in very low- and very high-yielding environments, the regular and DT corn hybrids did not show significant yield differences. In medium-yielding environments, DT corn hybrids were either equal to or had a yield advantage over regular hybrids most of the time. From these limited data, it was impossible to quantify the yield advantage of DT corn hybrids over regular hybrids, but information from the companies suggests that the new DT hybrids are expected to have a 10% yield advantage over regular dryland corn hybrids in specific yield environments. Assuming this is true at all corn yield levels, and assuming no significant change in drought tolerance of sorghum hybrids, the cutoff value for the relationship between dryland sorghum and corn declines to 5.4 Mg ha^{-1} (Figure 6.4). If drought tolerance increases yields by 50%, the cutoff yield decreases from 6 to 4.5 Mg ha^{-1}.

6.1.3 Land Use Efficiency by Region

The relationship between dryland corn and sorghum land use efficiencies in central, eastern, and western Kansas is presented in Figure 6.5. The majority of the data points from eastern Kansas are below the

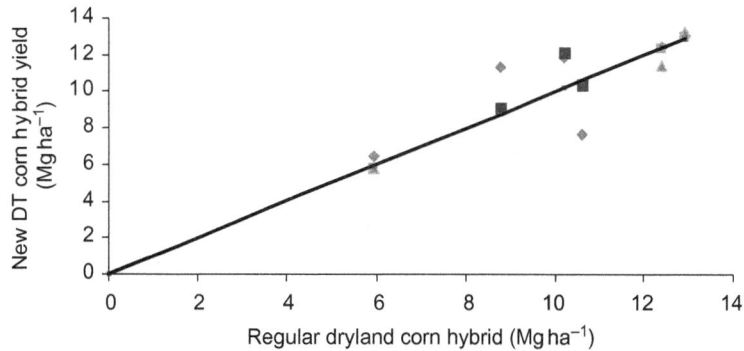

Figure 6.3 Relationship between regular corn hybrids and new DT corn hybrids from experiments at Etter, TX, and Topeka, KS, in 2011 and 2012. Solid line is a one-to-one relationship line; different size (colors) dots represent different DT corn hybrids.

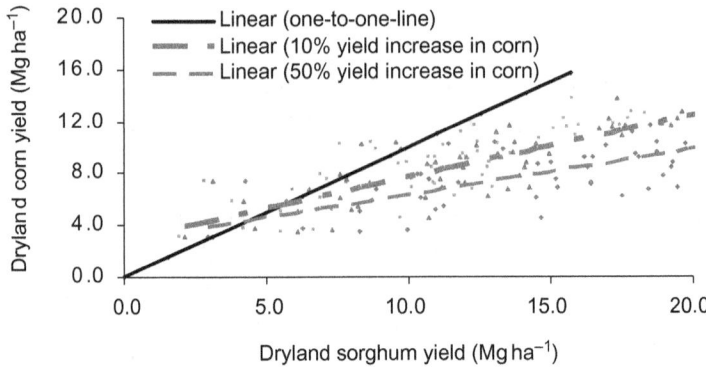

Figure 6.4 Relationship between dryland sorghum and actual and predicted corn yields based on Hybrid Performance Trial Data from 2007 to 2012 using hypothetical corn yield increases of 10% and 50% for DT hybrids over regular corn hybrids.

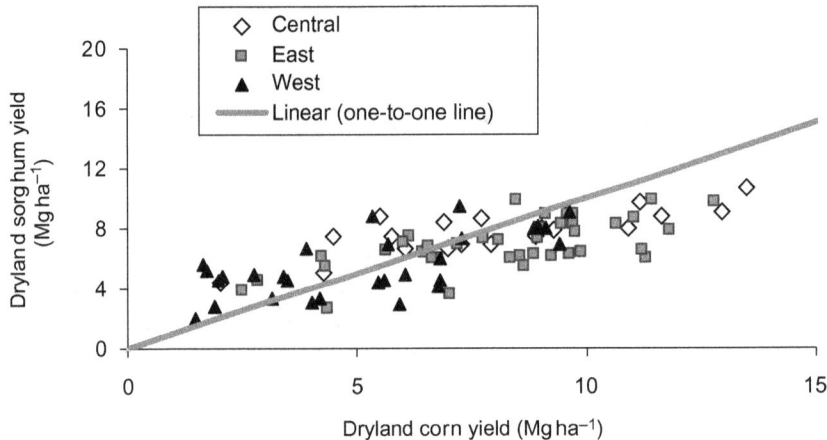

Figure 6.5 Relationship between dryland sorghum and corn yields in Kansas based on Hybrid Performance Trial Data in Kansas from 1992 to 2012 in different regions.

one-to-one line, whereas data points from central and western Kansas seem to be almost equally divided below and above the one-to-one line. On an average, dryland corn yield outyielded dryland sorghum 63% of the time in Kansas (Table 6.1), but this percentage varied depending on regions in Kansas. The probabilities that dryland corn outyielded dryland sorghum in different regions in Kansas are presented in Table 6.1. Based on our analysis, the probability that corn will outyield sorghum when corn yield is less than 6 Mg ha^{-1} is low (≤ 0.32). If expected yield is more than 6 Mg ha^{-1}, however, the probability that it will outyield sorghum is high (≥ 0.80).

Table 6.1 The Marginal and Conditional Probabilities of Dryland Corn Outyielding Dryland Sorghum in Different Regions in Kansas and in Different Productivity Scenarios

Region in Kansas	Probability that Dryland Corn will Outyield Dryland Sorghum	Probability that Dryland Corn will Outyield Dryland Sorghum, Given	
		Corn Yield <6 Mg ha^{-1}	Corn Yield >6 Mg ha^{-1}
East	0.77	0.13	0.94
Central	0.57	<0.01	0.80
West	0.48	0.32	0.80
Kansas	0.63	0.21	0.88

6.1.4 Land Use Efficiency Using Survey Data and Results from Other Studies

A comparison between corn and sorghum land use efficiency was conducted earlier by Staggenborg et al. (2008). An analysis similar to ours, Staggenborg et al. (2008), concluded that grain sorghum yielded better than corn in environments where corn yields were less than about 6 Mg ha^{-1}. Larson et al. (2001) conducted a trial comparing sorghum and corn in dryland, limited furrow, limited sprinkler, and full sprinkler irrigation conditions. The results of their experiment indicated that sorghum outyielded corn only in dryland production in southeastern Colorado, where reported yields of both crops were below 6 Mg ha^{-1}; however, corn outyielded sorghum with limited and full irrigation practices where reported yields of corn were greater than 6 Mg ha^{-1}.

USDA National Agricultural Statistic Service (NASS) reports the yield of corn and sorghum annually for each county in the United States. The relationship between corn and sorghum yield based on these data is given in Figure 3.8. The cutoff values calculated from the data, 4 Mg ha^{-1}, were significantly lower than the cutoff values derived from hybrid performance trial data. The reported corn and sorghum yields in the USDA dataset likely are not from the same or adjacent fields, making crop yield comparisons suspect given potential differences in soil depth, texture, cropping history, residue cover, soil profile water, and so on. Therefore, the potential for significant confounding of crop with local field environment in this dataset may be the cause of the discrepancy in cutoff value between the two datasets.

6.2 WATER REQUIREMENTS AND WATER USE EFFICIENCY

6.2.1 General Water Requirements and Water Use Efficiency

The maximum water requirements of full-season corn and sorghum plants are about 635 and 533 mm, respectively (Stone et al., 2006). Based on the same paper by Stone and Schlegel (2006), the relationships between corn and grain sorghum yield and evapotranspiration (ET) are given in Figure 6.6A. A minimum of about 277 and 175 mm of water is required by corn and sorghum, respectively, to support vegetative development alone. For each additional 25 mm of water above this threshold, on an average, corn yields about 1.06 Mg ha^{-1} and sorghum yields about 0.77 Mg ha^{-1}. Figure 6.6B presents the relationship between WUE of corn and sorghum at different ET values. WUE was calculated by dividing yield by cumulative ET. At ET values less than 537 mm, sorghum had better WUE, and at ET greater than 537 mm, corn had better WUE.

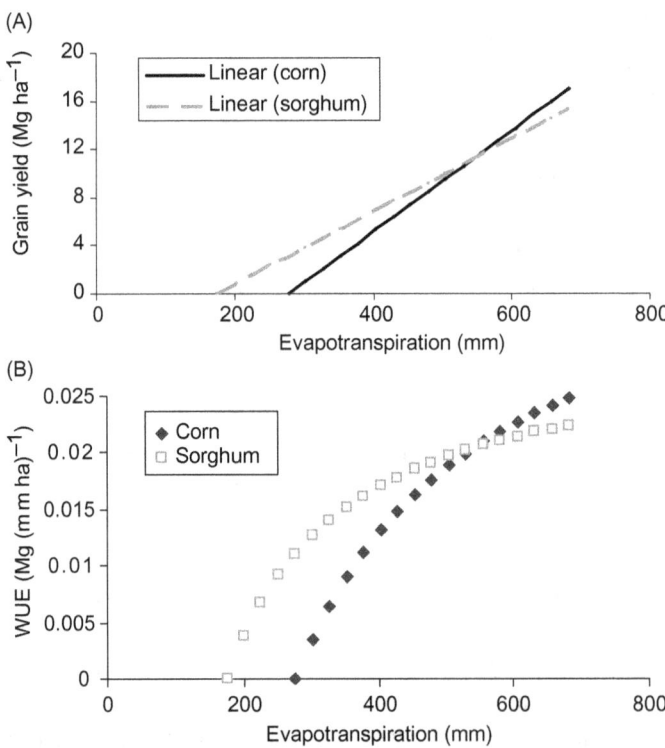

Figure 6.6 Water requirement versus yield (A) and water use efficiency (WUE) (B) based on the data from Stone et al. (2006).

6.2.2 Rainfall Use Efficiency

In the above section, we considered water and yield relationship accounting for losses such as runoff, evaporation, and deep percolation and stored soil from previous seasons. In this section, we consider the general rainfall–yield relationship without accounting for stored soil from the previous season to delineate the seasonal rainfall cutoff point for corn or sorghum production. Based on data from Kansas State University Hybrid Performance Trials (1992–2012), the relationship between dryland yield and total seasonal rainfall (April to September) is presented in Figure 6.7A. The relationship between rainfall use efficiency (defined as yield/season rainfall) and seasonal rainfall is presented in Figure 6.7B. Based on this analysis, dryland sorghum outyielded and had better rainfall use efficiency than dryland corn for seasonal rainfall amounts less than about 432 mm. On the other hand, dryland corn had better yield and rainfall use efficiency in places or years where seasonal rainfall was greater than about 432 mm.

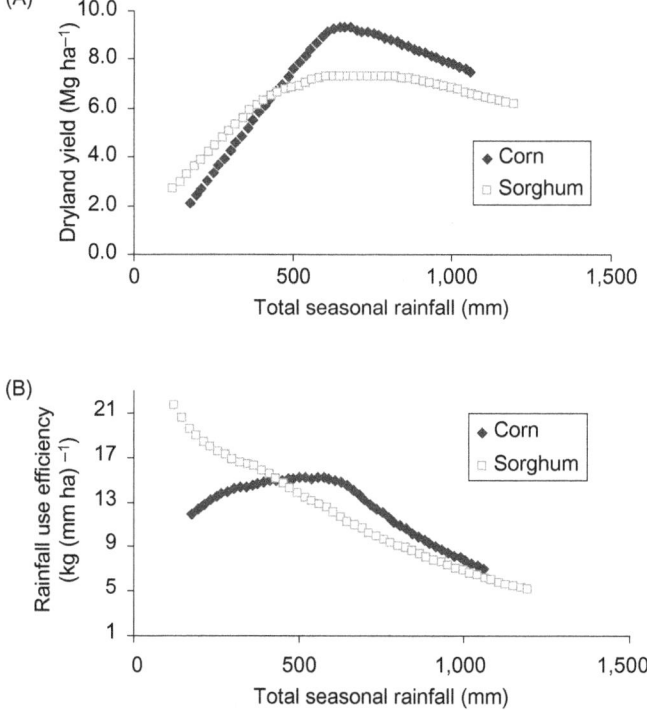

Figure 6.7 Total seasonal rainfall (April to September) versus yield (A) and rainfall use efficiency (B) based on yield and precipitation data from Hybrid Performance Trials in Kansas from 1992 to 2012.

6.2.3 Other Studies on WUE

A study by Rees and Irmak (2012) compared crop water use of corn, sorghum, and soybean in Nebraska from 2009 to 2011. They reported crop WUEs of 6.7, 4.3, and 5.8 bushels for corn and 5.6, 5.5, and 8.0 bushels (acre-inch)$^{-1}$ for sorghum for the years 2009, 2010, and 2011, respectively. In 2009, corn had better WUE, which was attributed to lack of rain to activate sorghum herbicide and resulting high grassy weed pressure. Overall, the investigators concluded that sorghum uses water more efficiently but indicated that their study did not include new "DT" corn hybrids.

A study by Klocke and Currie (2009) in Garden City, KS (Figure 6.8A), provided additional information about the relationship between ET and yields of corn and sorghum. It can be deduced that with more than 483 mm of ET, corn was more efficient than grain sorghum. Figure 6.8B shows the relationship between water supply and

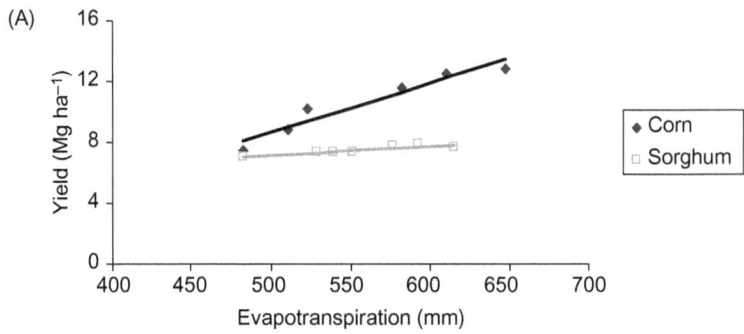

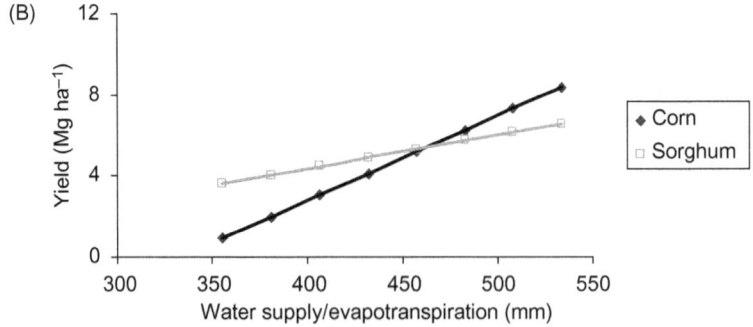

Figure 6.8 (A) Relationship between ET and yield of corn and sorghum based on data from Klocke and Currie (2009) and (B) based on yield–water supply equations given for sorghum by Stone and Schlegel (2006) and yield and ET for corn from Klocke et al. (2011).

sorghum yield based on an equation developed by Stone and Schlegel (2006) and the relationship between ET and corn yield based on data from Klocke et al. (2011). The two lines intersected at about 457 mm of ET, suggesting this as a cutoff value for efficiency of corn and sorghum.

Stone et al. (1996) reported that corn outyielded sorghum at total irrigation amounts of 356 mm and above at Tribune, KS; however, the gain in corn yield with additional irrigation was greater for corn starting at 203 mm of irrigation water. Therefore, their report recommended 203 mm of irrigation water as a cutoff below which sorghum is more efficient and above which corn is more efficient. Tribune's normal seasonal (April to September) rainfall is about 330 mm. This sums to a total of 533, which is close to the 537 mm cutoff value presented above for the ET−yield relationship.

6.3 FERTILIZER REQUIREMENTS AND FERTILIZER USE EFFICIENCY

6.3.1 Nitrogen Use from K-State Fertilizer Recommendation

The fertilizer requirements of crops are highly dependent on the soil test value of the nutrient in question, the amount of soil organic matter, and the yield goal, among other factors. Figure 6.9 presents fertilizer recommendations for different yield goals for corn and sorghum based on Kansas State University's soil test interpretation and fertilizer recommendation booklet (Leikam et al., 2003). Based on this

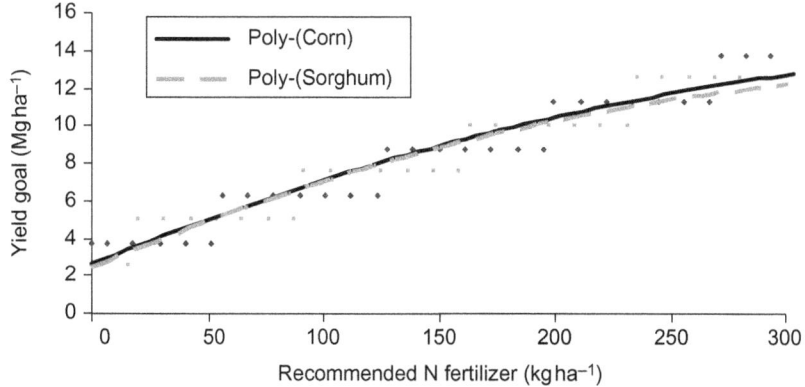

Figure 6.9 Recommended N fertilizer levels at various yield goals and soil organic matter levels for corn and sorghum. Data are from Kansas State University soil test interpretation and fertilizer recommendation booklet.

relationship, corn and sorghum have a similar N fertilizer recommendation and use efficiency from 0 to about 224 kg ha^{-1}, above which corn is more efficient.

6.3.2 Nitrogen Fertilizer Use and N Use Efficiency Based on Hybrid Trials

The relationships between applied N fertilizer with yield and N use efficiency (NUE) for corn and sorghum from Hybrid Performance Trials from 1992 to 2012 are presented in Figure 6.10. In Figure 6.10A, the yield responses of corn and sorghum at different rates of applied fertilizer are presented. From this dataset, we can deduce that corn responded more than grain sorghum to N fertilizer applications of 50–200 kg ha^{-1}. In Figure 6.9B, the NUEs of the two crops are presented. For equal rates of N fertilizer application, corn responded about 69 kg (ha kg N)$^{-1}$ more than sorghum.

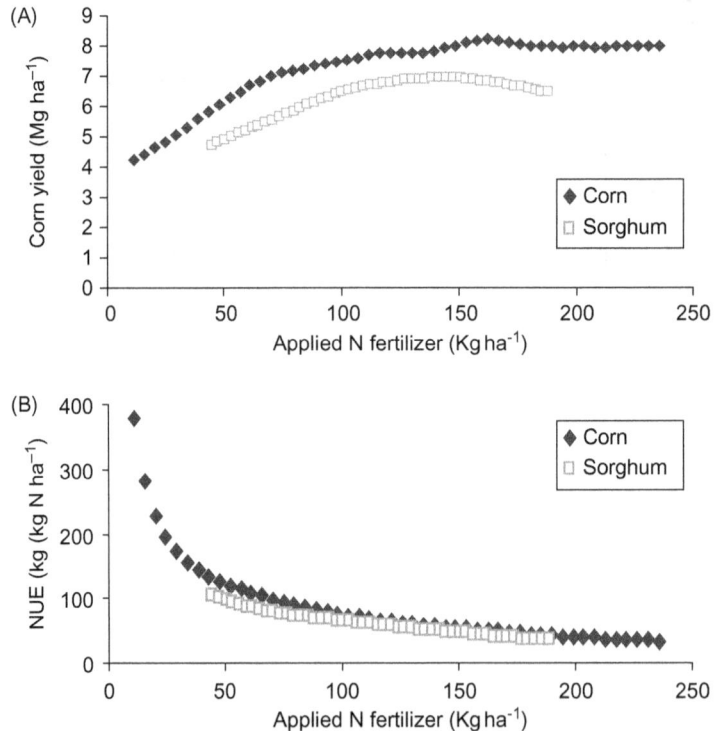

Figure 6.10 (A) Relationship between applied N fertilizer levels with yields and (B) with NUE of corn and sorghum in Hybrid Performance Trials from 1992 to 2012.

6.3.3 Other Studies on Nutrient Use Efficiency

Results from a long-term (1997–2006) N and phosphorus (P) fertilizer use study for corn and sorghum from Schlegel (2006a,b) are presented in Figure 6.11. The research was conducted in Tribune, KS. The trend lines indicate that corn responds to applied N and P more than sorghum at the considered application levels; however, it should be noted that corn yield varies more than sorghum yield, perhaps due to different levels of rainfall to supplement irrigation in the years this research was conducted.

6.4 PESTICIDE REQUIREMENTS AND USE EFFICIENCY

The relationship between dryland corn and sorghum yields with and without pre- and postemergent herbicide application is presented in

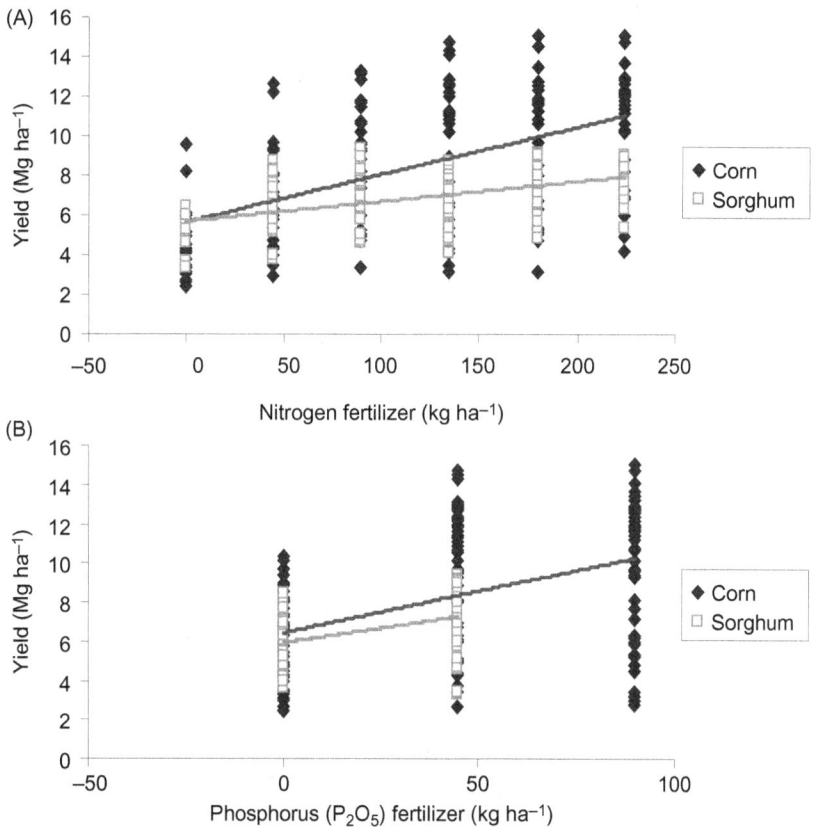

Figure 6.11 Relationship between applied nitrogen fertilizer levels (A) and applied phosphorus fertilizer (B) with corn and sorghum yields based on long-term trials in Tribune, KS.

Figure 6.12. These data come from herbicide trials conducted in the years 2008, 2009, 2011, and 2012 in Manhattan, KS. Yield response to herbicide application was different in each year. With heavy weed pressure, both corn and sorghum benefit from a pre- followed by postemergence herbicide application, as shown in 2008. Lower sorghum yield from postemergence-only applied herbicide in 2009 and 2011 suggests the presence of annual grassy weeds in the experiment that were left uncontrolled.

On an average, herbicide application resulted in about 40–100% yield increases in corn and sorghum compared with the untreated check (Figure 6.12 summary). Postemergent herbicide application on dryland corn was found to be as effective as or better than preemergent herbicides because no glyphosate-resistant weeds were present in the experiment. On the other hand, dryland sorghum yields with only a

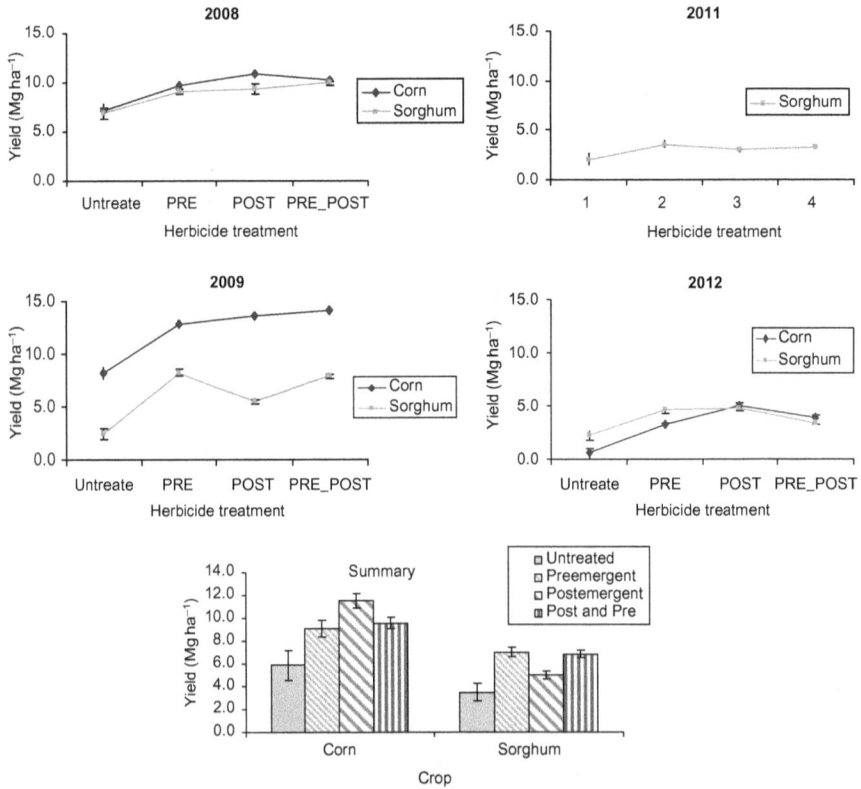

Figure 6.12 Relationship between timing of herbicide application and yield of dryland corn and sorghum based on research in Manhattan, KS (2008–2012), and with summary bar graph.

postemergent herbicide application were lower than those with a preemergent application.

Reasonable weed control can be accomplished in sorghum and corn using preemergent herbicide application, but preemergent herbicides require adequate rainfall to be activated and control weeds. If rainfall is inadequate to activate these herbicides, weeds might escape and a postemergent herbicide might be required. Because sorghum is not glyphosate-tolerant, the options for postemergence herbicides, particularly for grass weeds, are much less effective than postemergent herbicide options available in corn (Kershner et al., 2012; Thompson et al., 2009), which implies that corn is better than sorghum in herbicide use efficiency, assuming herbicide rates for these crops are similar.

6.5 PRELIMINARY ECONOMIC CONSIDERATIONS (GENERAL RESOURCE USE EFFICIENCY)

Costs of production and prices of grain sorghum and corn are different, so the cutoff values we suggested above should not be used independently for decision making. Figure 6.13 contains cutoff values based on the relationships presented above with inclusion of a net revenue consideration. In order to determine the simple economic cutoff value, the yield relationship between corn and sorghum in Figure 6.1A is used. The yield of corn and sorghum from 0 to 20 Mg ha^{-1} was simulated based on the relationship in Figure 6.1. At each yield value, the

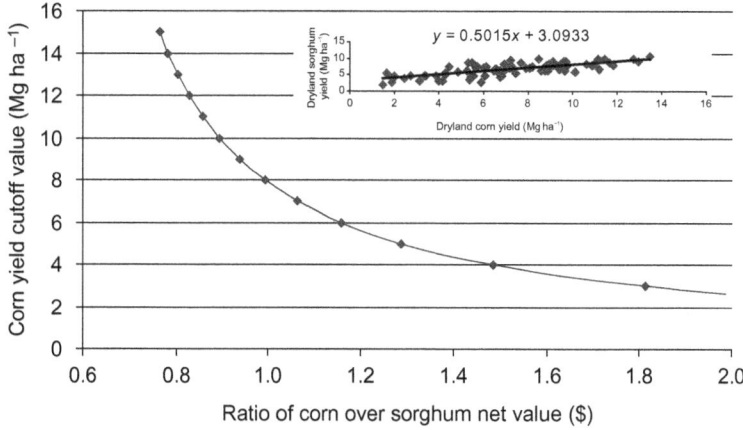

Figure 6.13 *Cutoff corn yield values based on dryland sorghum−corn yield relationship and ratio of corn to sorghum net return.*

net return of corn (NRC) that equalizes it with the net return of sorghum (NRS) was calculated in Eq. (6.1). The ratio of this NRC to NRS value is constant at different values of NRS; therefore, we can use the NRC to NRS ratio at different yield values of corn to determine cutoff values as outlined in the steps of Eqs. (6.1)–(6.3).

At the economic cutoff value, we have:

$$(CY)(NRC) - (SY)(NRS) = 0 \qquad (6.1)$$

where CY and SY are corn and sorghum yields, respectively.

Corn and sorghum relationship based on Figure 6.1A is

$$SY = 0.5015(CY) + 3.0933 \qquad (6.2)$$

From Eqs. (5.1) and (6.1), we can calculate corn yield as function of ratio of net return as:

$$CY = \frac{3.0933}{NRC - 0.5015(NRS)} = \frac{3.0933}{(NRC/NRS) - 0.5015} = \frac{3.0933}{RatioNR - 0.5015} \qquad (6.3)$$

As shown in the Figure 6.13 when the ratio of the NRC over NRS is 1, the cutoff value below which sorghum and above which corn is relatively more efficient is about 6.2 Mg ha^{-1} of corn. If the ratio is greater than 1, the cutoff value will be less than 6.2 Mg ha^{-1} of corn yield, and if the ratio is greater than 1, it is greater than 6.2 Mg ha^{-1} of corn yield. The relationship illustrated in the graph or the equation above could be used to determine the cutoff value as market price fluctuates.

CHAPTER 7

Rotation Effects of Corn and Sorghum in Cropping Systems

Crop rotation is one of the earliest and most important crop management systems. Rotation of crops, as opposed to continuous cropping, contributes to crop yield and health because of improvements in soil structure (Barber, 1972), improvement in plant nutrition due to differences in the feeding range of roots (Peterson and Varvel, 1989), enhancement of soil moisture (Benson, 1985), disruption of insect and disease cycles, and improved weed control (Slife, 1976; Dick and Vandoren, 1985; Benson, 1985).

The most common rotation system in the Great Plains before improvements in fertilizer, pesticide, and reduced/no-till technologies was a 2-year wheat-summer fallow system (Anderson et al., 1999; Hansen et al., 2012; Norwood, 2000). Improvements in the aforementioned technologies enabled including summer crops such as sorghum, corn, soybean, and sunflower in rotations with winter wheat. A rotation of a winter crop followed by a summer crop improves land and water use (Nielsen et al., 2002; Hansen et al., 2012), but every rotation system has advantages and disadvantages. These advantages and disadvantages can be characterized by analyzing the compatibility between crops within a rotation in terms of planting and harvest date requirements, water use, and yield.

The main objectives of our analysis were to identify the most common rotation systems in the Great Plains that involve corn and/or sorghum and to compare the advantages and disadvantages of corn and sorghum in these rotations. The data for this comparison were assembled from a survey conducted in 2007, from results of rotation studies conducted in different parts of the Great Plains, and from crop water use, planting and harvest dates, and long-term weather information.

7.1 ANALYTICAL APPROACH

To compare the performance of corn and sorghum crop rotations, data were assembled from several research studies. The first dataset

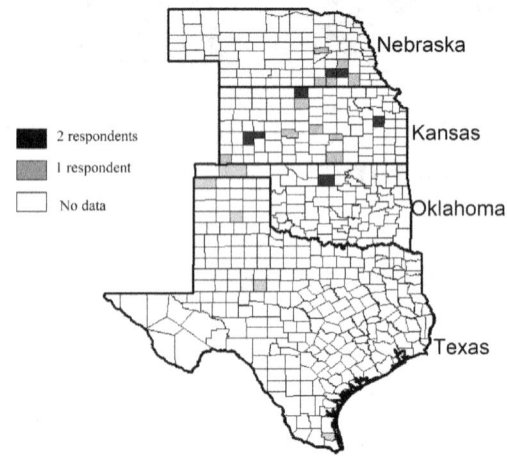

Figure 7.1 Map depicting number and distribution of respondents to 2007 survey of cropping practices.

was from a survey of dryland farmers regarding their common crop rotations conducted by the Cropping Systems Research Project in the Kansas State University's Department of Agronomy in 2007. An open-ended survey was developed that questioned farmers about the crops that they grew, their crop rotations, how they made decisions about crop sequence, what crop sequences they found more beneficial, and the most probable reason for the benefit. The survey was sent to different farmers via email, and responses were collected by mail, email, or fax. About 28 farmers responded to the survey. Figure 7.1 presents the number and distribution of respondents.

Even though the number of respondent farmers to the survey was small compared with the area covered, multiple responses about rotation systems often were received from a single farmer. Depending on the year and conditions in different parts of their farm, farmers reported using different rotations within a year or in different years. This makes the number of responses more than twice the number of respondents. To capture spatial differences in the most common rotation systems in the region, the data from Kansas and Nebraska, which had greater sample sizes, were analyzed separately. Additional information collected with this survey was used to establish some of the most important factors farmers use to determine their crop rotations. Farmer responses also provided an indication of their opinions regarding the performance of corn and sorghum in their crop rotations.

Table 7.1 Rotation Studies Including Wheat, Corn, Sorghum, or Soybean Conducted at Different Locations in the Great Plains

Author	Year	Location	Crops in Rotation	Other Variables
Claassen	2003	Hesston, KS	W,GS	Fertilizer
Claassen	2006	Hesston, KS	W,C,GS,SB	Tillage
Claassen and Regehr	2009	Hesston, KS	W,GS,SB,C,SF	–
Gordon	1991	Belleville, KS	W,C	Fertilizer
Halvorson et al.	2004	Akron, CO	W,C,GS,F	Fertilizer
Heer	1999	Hutchinson, KS	W,C,SB,GS	Fertilizer
Heer	2009	Hutchinson, KS	W,GS	Fertilizer
Kelley and Sweeney	2009	Parsons, KS	W,GS,SB,C	Fertilizer
Norwood	1999	Garden City, KS	W,C,GS,SB,SF	Tillage
Schlegel	1994	Tribune, KS	W,C,GS	None
Schlegel and Frickel	1994	Tribune, KS	W,GS,F	Fertilizer
Schlegel et al.	2010	Tribune, KS	W,GS,SF,F	Tillage
Schlegel et al.	1994	Tribune, KS	C,GS,SB	Fertilizer
Tarkalson et al.	2006a	North Platte, NE	C,GS,F	Tillage
Wary et al.	1994	Cherokee, KS	W,C,GS,SB	Fertilizer

W, wheat; GS, grain sorghum; SB, soybean; SF, sunflower; C, corn; F, fallow.

The second dataset was assembled from 15 separate rotation studies conducted in the Great Plains region. Table 7.1 summarizes who reported these studies and where and when they were conducted. These studies were grouped into 2-, 3-, and 4-year rotations. The effects of corn and sorghum on yield of a following wheat or soybean crop were compared in equivalent rotation systems. Mean wheat or soybean yield for the entire period of the study was assembled from each study. In most cases, mean yields were not only across years but were collected for different fertilizers or tillage treatment levels of the study.

Data from the assembled studies were analyzed using PROC MIXED in SAS. Wheat or soybean yield was modeled as a function of rotation (fixed effect) with other variables (year, fertilizer, or tillage) treated as random effects based on the assumption that these covariates represented random samplings of all possible years and management effects on the rotation systems under consideration. Results were summarized from studies because the other variables, i.e., years the study was conducted, fertilizer treatments, or types of tillage, differed from one study to another. Studies that reported equivalent rotation systems are presented side by side to help visualize the effects of corn or sorghum in crop rotations across studies or locations.

Finally, planting and harvest date information, crop water use, and freeze dates were assembled for wheat, corn, and sorghum crops. Based on this information, compatibility of wheat planting after corn and sorghum harvest as well as double-cropping those crops after wheat were evaluated.

7.2 SURVEY OF COMMON ROTATIONS AND FACTORS AFFECTING CROP SEQUENCING DECISIONS

More than 50% of respondents from Kansas indicated that they grew sorghum, corn, wheat, soybean, or sunflower. Less than 20% of the Kansas respondents indicated that they grew alfalfa, canola, sudangrass, or rye. The two most common rotations reported in Kansas were (i) wheat followed by a summer crop, which was then followed by fallow (W-SC-F), and (ii) 2 years of wheat followed by a summer crop, which was followed by another summer crop (2W-2SC) (Figure 7.2). In the W-SC-F system, the most common summer crop components were sorghum, corn, or sunflower. In the 2W-2SC rotation, the first summer crop component was usually either corn or sorghum, and the second summer crop component was soybean or sunflower. Based on these responses, we can generalize that wheat (1 or 2 years) followed by corn, sorghum, or sunflower, which was then followed by fallow or soybean was the major crop rotation of respondent farmers from Kansas (Figure 7.2).

Kansas farmers were asked what cropping sequence they found most beneficial. They indicated that sorghum after wheat was a reliable rotation in most years. Farmers stated that corn after wheat worked in good years, but wheat after wheat was a better alternative than sorghum after wheat in dry years. Respondents indicated that in a W-C/GS-SB rotation system, the intention of rotating soybean after corn or sorghum is to plant wheat after soybean harvest in the fall. Farmers also said following soybean with corn or sorghum was beneficial due to fixed nitrogen (N) and interrupted weed propagation. One farmer indicated that any crop after canola performed better than after other crops. Another farmer indicated that wheat after wheat for 3–4 years and sorghum after sorghum for 3 years proved stable in dry

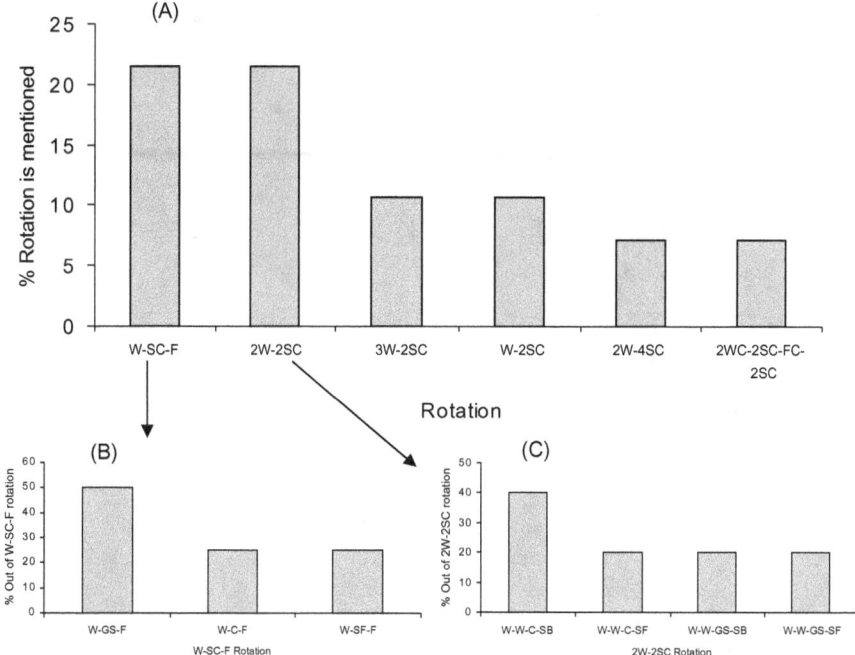

Figure 7.2 Rotation systems in Kansas that represent at least 5% of the systems mentioned by respondent farmers (A) and rotation systems under the most frequently mentioned W-SC-F (B) and 2W-2SC (C). W, wheat; SC, summer crop; WC, winter crop; FC, forage crop; GS, grain sorghum; C, corn; F, fallow; SF, sunflower; SB, soybean; 2W, W-W; 2SC, SC-SC.

areas with minimal weed pressure. Dryland farmers planted sunflower when the price was attractive and seasonal rainfall appeared favorable. Sunflower was believed to tap deeper moisture than sorghum and did not lodge as badly as sorghum when stressed by drought.

In Nebraska, respondent farmers commonly grew corn, sorghum, wheat, and alfalfa. Compared with Kansas, fewer respondents included wheat in their rotation. The most common rotation system mentioned by Nebraska respondents was a 3-year rotation of the summer crops corn, sorghum, and soybean (C-GS-SB), i.e., corn followed by sorghum, which was then followed by soybean (Figure 7.3). Other rotation systems found in relatively higher frequency included a 2-year rotation of corn and soybean (C-SB) and wheat followed by grain sorghum then soybean (W-GS-SB).

When they responded to the question of which crop sequence they found more beneficial and the possible reason, Nebraska respondents suggested sorghum rather than corn after wheat for accessing moisture

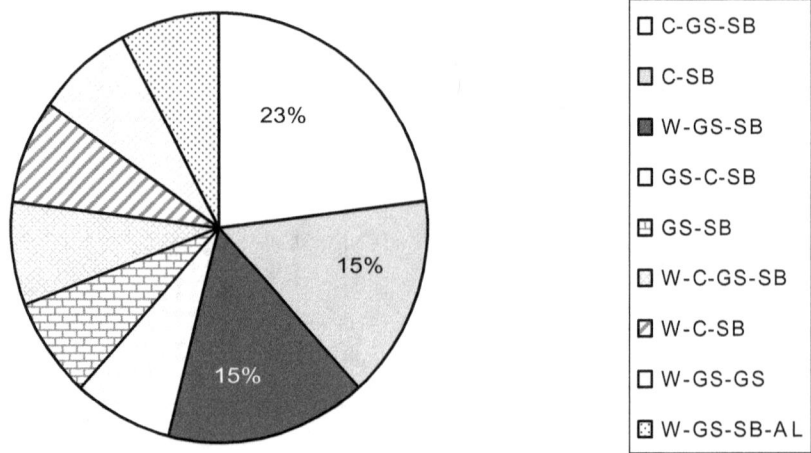

Figure 7.3 Frequency of rotation systems for Nebraska respondents as a percentage of all rotation systems mentioned by respondent farmers. W, wheat; GS, grain sorghum; C, corn; SB, soybean; AL, alfalfa.

in drought conditions. Sorghum or corn after soybean or alfalfa was also preferred due to N supply. Corn following sorghum was also suggested as beneficial due to snow capture by sorghum stubble. Farmers also indicated that they chose corn rather than sorghum followed by soybean for weed control because they felt that sorghum herbicides are not as effective as glyphosate for weed control.

Survey responses indicated that crop selection and sequencing (rotation) was dictated largely by soil moisture status or irrigation capacity and weed and other pest pressure (pesticide options) (Figure 7.4); however, factors such as value (price) of crop and inputs, expected risk, planting date, and previous crop were also important to dictate cropping sequence of farmers. Corn was preferred over sorghum in good years for its potential for better yield, better herbicide choices for a following crop, and better market price. In years with moisture deficit, sorghum was suggested as a better choice than corn. In cases where farmers wanted to rotate from corn or sorghum to wheat, it was suggested that corn left more water for the following wheat.

These findings regarding the common rotation systems in the Great Plains region agree with previous reports. The W-GS/C-F rotation has been reported to account for 90% of the crop area in some counties of Nebraska and Kansas and was described as the common rotation system in the Great Plain region (Tarkalson et al., 2006a,b; Wicks et al.,

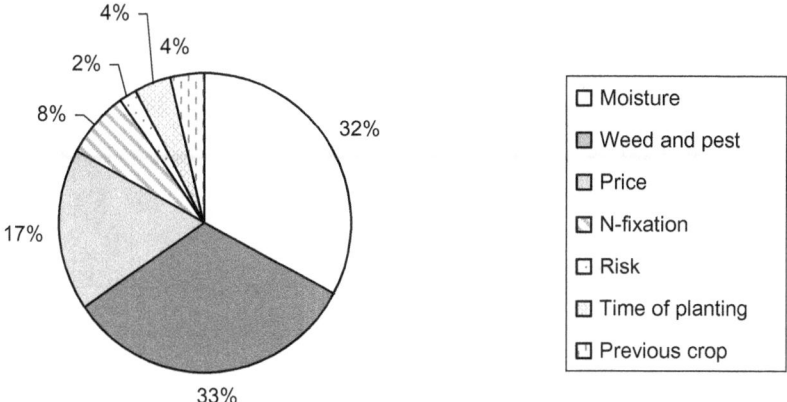

Figure 7.4 Factors that dictate rotation system decisions by percentage mentioned in survey respondents from Kansas, Nebraska, Texas, and Oklahoma.

1995; Hansen et al., 2012). The advantages and possible consequences of these rotations pertaining to land use, moisture, and pest and weed control were also discussed in these papers.

7.3 ANALYSIS OF ROTATION STUDIES

7.3.1 Corn and Sorghum in 2-Year Rotation with Wheat and Soybean

An assembly of 15 separate rotation studies in the Great Plains facilitated comparing the effects of corn and sorghum on winter wheat and soybean yield in different rotation systems (Table 7.1). The impact of corn and sorghum on yield of the following wheat and soybean crop in a 2-year rotation was studied by analyzing data obtained from four studies out of the 15 using a mixed model (Figure 7.5). Three of the four studies compared wheat yield in a 2-year rotation: a 1-year study in southeast Kansas reported by Wary et al. (1994); a multiyear rotation study in southeast Kansas with different fertilizer levels from Kelley and Sweeney (2009); and another multiyear study with a tillage comparison from south-central Kansas reported by Claassen (2006). The fourth study, reported by Schlegel et al. (1994), was used to compare the effects of corn and sorghum in a 2-year rotation with soybean.

The analysis of the three studies showed no significant difference in wheat yield after corn or sorghum, except in the study by Kelley and

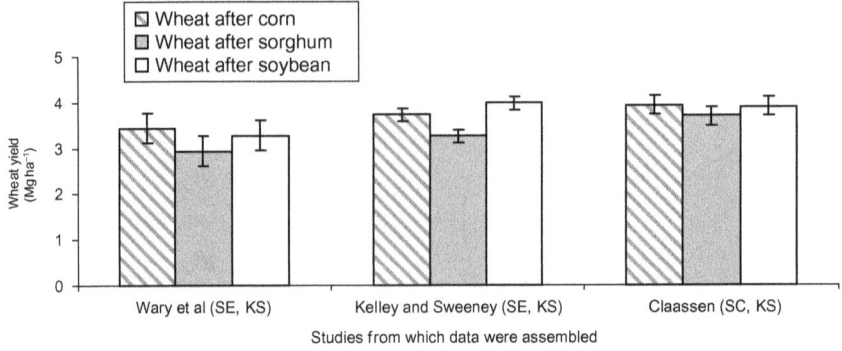

Figure 7.5 Wheat yield after corn, sorghum, and soybean crops in a 2-year rotation.

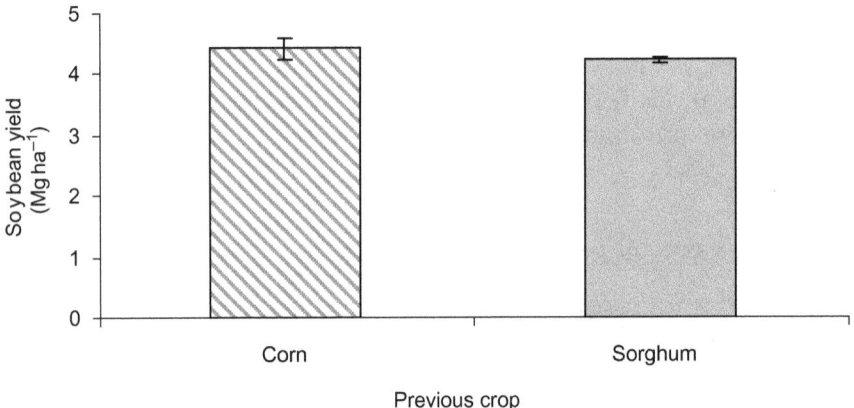

Figure 7.6 Soybean yield after corn and sorghum in Garden City, KS, from 1991 to 1993 (Schlegal et al., 1994).

Sweeney (2009), although mean yield of wheat following sorghum was always the least (Figure 7.5). When the data obtained from Kelley and Sweeney were analyzed for the effects of rotation (fixed factor) with years and fertilizer levels as random effects, wheat yields after corn were greater than after sorghum. In the other two studies, wheat yields did not differ after corn compared with wheat yields after sorghum. Similarly, yields of soybean were about 2 bushel acre^{-1} greater but more variable after corn than after sorghum in an irrigated study at Garden City (Figure 7.6).

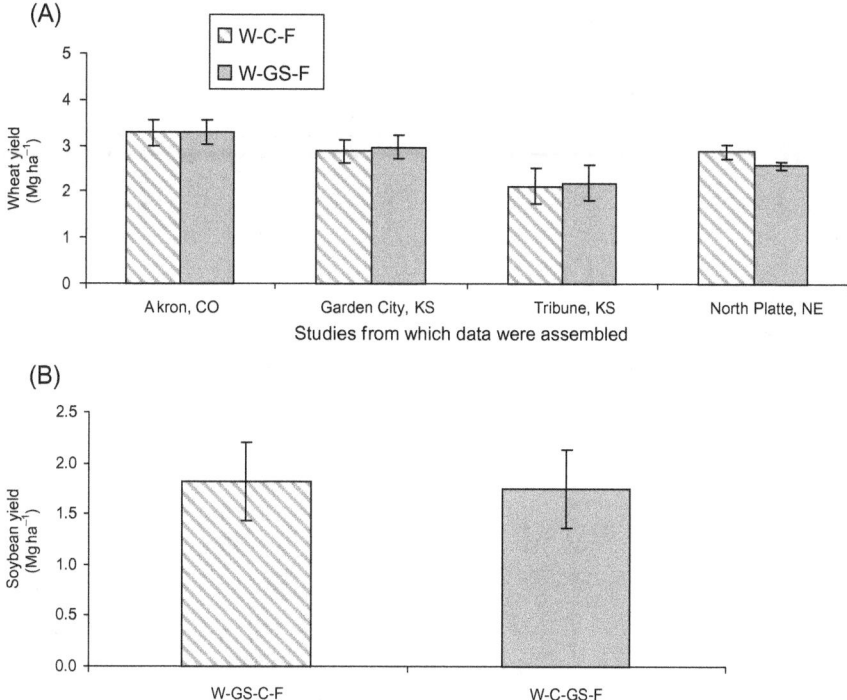

Figure 7.7 Wheat yield after corn and sorghum in a wheat-summer crop-fallow (W-SC-F) or wheat-summer crop-summer crop-fallow (W-SC-SC-F) rotation systems.

7.3.2 Corn and Sorghum in a 3- or 4-Year Wheat-Summer Crop-Fallow System

The effects of corn and sorghum on wheat yield in a wheat-summer crop-fallow (W-SC-F) or a wheat-summer crop-summer crop-fallow (W-SC-SC-F) system was studied by extracting multiple-year data from 5 of the 15 studies (Figure 7.7). The four W-SC-F studies were assembled from Colorado, Nebraska, and two studies from western Kansas (Table 7.1 and Figure 7.7A). Similar to the 2-year rotation systems, only one (North Platte, NE) of the four studies showed a greater yield of wheat in rotation with corn compared with rotations with sorghum in a W-SC-F system. In that particular study (Tarkalson et al. 2006a), however, the years when the W-C-F and W-GS-F research was conducted were different, so rotation and environment (year) effects were confounded. Wheat yield did not differ significantly between the W-GS-C-F and W-C-GS-F rotation systems.

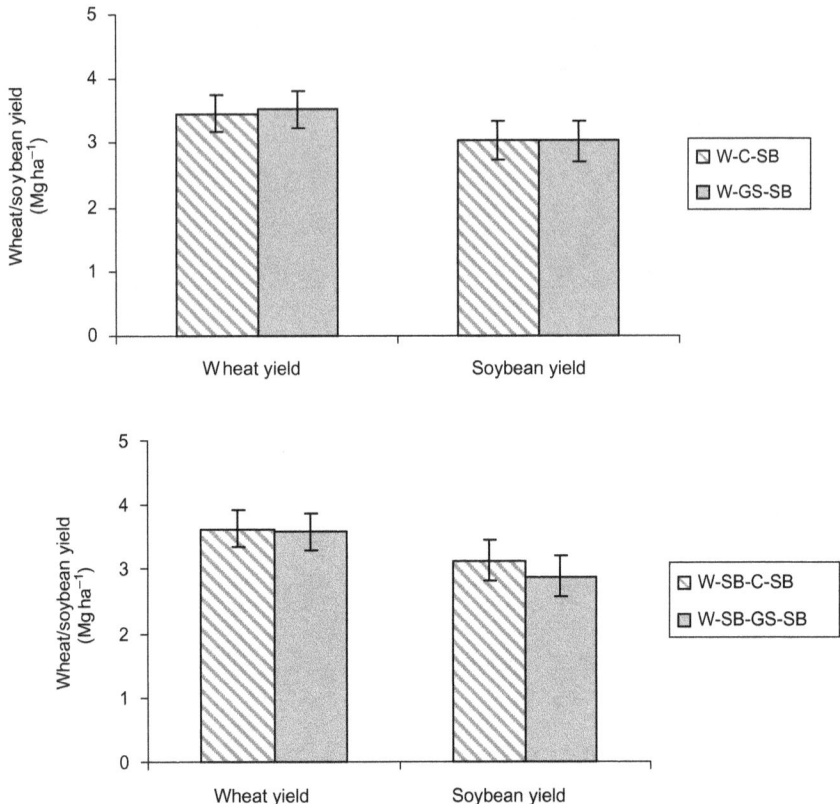

Figure 7.8 Wheat and soybean yield in a 3-year W-C/GS-SB and W-SB-C/GS-SB rotation systems.

7.3.3 Corn and Sorghum in 3- or 4-Year Rotation with Both Wheat and Soybean

The impact of corn and sorghum on both wheat and soybean in a 3- or 4-year rotation was analyzed from a multiyear no-till rotation study reported by Claassen and Regehr (2009), who studied 11 different crop rotation systems, four of which provided comparisons relevant to our objective. Figure 7.8 presents wheat and soybean yields in a 3-year wheat-corn/sorghum-soybean (W-C/GS-SB) and a 4-year W-SB-C/GS-SB rotation system. Neither wheat nor soybean yields differed in equivalent corn and sorghum rotations in this study.

7.3.4 Fertilizer Requirements of Wheat After Corn and Sorghum

Evaluations of wheat yield in rotation with sorghum or corn with different rates of N fertilizer were reported in several of the assembled

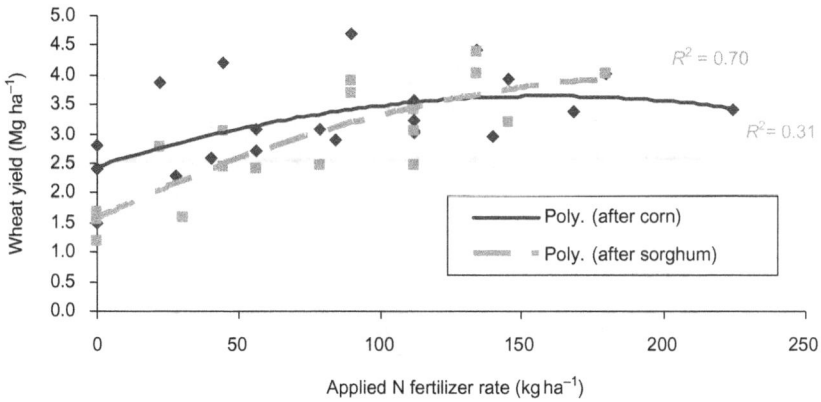

Figure 7.9 Wheat yield response to N fertilizer after corn and sorghum.

studies (Table 7.1). Analysis of the effects of fertilizer on wheat yield after corn and sorghum is presented in Figure 7.9. These combined results show that wheat yields after corn were greater than after sorghum when N application to the wheat crop was less than about 100 kg ha^{-1}. Wheat yields did not differ after corn compared with sorghum when N application was more than 100 kg ha^{-1}.

These results are consistent with findings of others who reported higher N requirements for wheat after sorghum than after wheat, soybean, or corn (Knowles et al., 1993; Staggenborg et al., 2003; Wary et al., 1994). Possible reasons given for the greater N requirement after sorghum is that wheat planted after sorghum will likely be later than optimum, and the length of time between sorghum harvest and wheat planting is shorter than after corn. This may result in a decrease in the tillering potential of wheat. Other possible reasons could be greater N immobilization potential of sorghum residue compared with corn residue or allelopathic effects of sorghum on the following wheat crop (Knowles et al., 1993; Roth et al., 2000; Schmidt and Frey, 1988).

7.4 ANALYSIS OF CROP SEQUENCING RELATIVE TO CROP WATER USE AND OPTIMUM PLANTING AND HARVEST DATES

The optimal planting and harvesting dates of crops in a rotation system is an important compatibility factor for a rotation system analysis. Here, we used the recommended or common planting and harvesting

98 Corn and Grain Sorghum Comparison

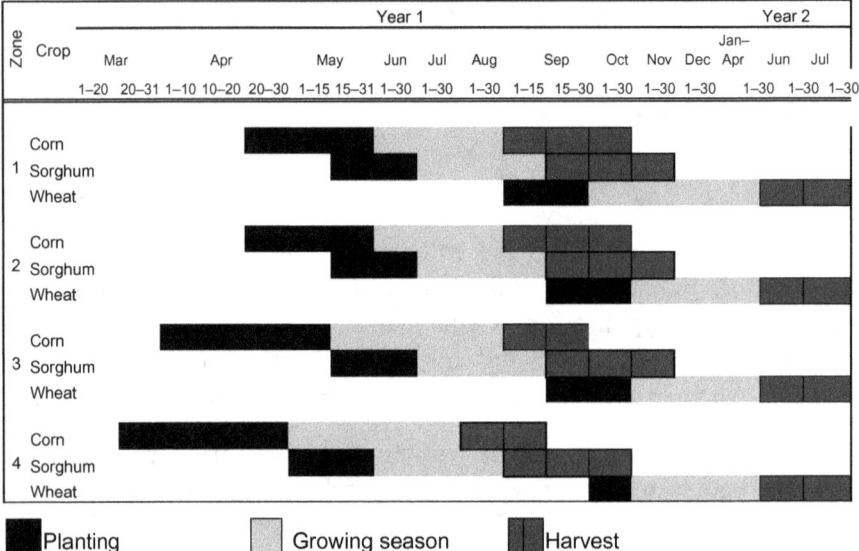

Figure 7.10 Map of Kansas with different zones, adapted from Shroyer et al. (1996), and a chart depicting recommended planting and harvesting dates for corn, sorghum, and wheat.

dates of wheat, corn, and sorghum in Kansas to analyze fitness of corn and sorghum for rotation with wheat (Figure 7.10).

The recommended planting dates for wheat, corn, and sorghum in Kansas vary from one zone to another (Figure 7.10 and Shroyer et al., 1996). In zone 1 (northwest Kansas) and zone 2, recommended wheat planting dates range from the second week of September to the end of October. Harvest usually occurs from mid-June to the end of July. Corn planting dates for these two zones are from the second week of April to the third week of May. Corn is typically harvested from the first week of September to the end of October. On the other hand, the recommended sorghum planting date in these zones is late May to the end of June, and harvest is from late September to the end of November. If wheat must be planted after sorghum, in the worst case, planting will be pushed into November, 2 months later than the recommendation. If wheat is planted after the late-harvested corn, planting is only 1 month later than the recommendation. Common practices and crop management recommendations reveal a much larger overlap for wheat after sorghum compared with after corn in zones 1 and 2.

In zones 3 and 4 (southeast half of Kansas), typical wheat planting dates ranged from late September to the end of October. Harvest was from mid-June to mid-July. Corn planting dates for these two zones begin from the last week of March and continues through the first week of May. Corn is typically harvested from as early as the second week of August and is complete by mid-September. The recommended sorghum planting date in these zones is from early May to the end of June. Harvest typically is from early September to the end of October. If wheat is to be planted after sorghum in zones 3 and 4, planting will be pushed until the end of September, which is close to the recommended planting time. Little or no overlap occurs between wheat planting and corn harvest in zones 3 and 4.

If we look at corn or sorghum planting after wheat, the above story is reversed. In general, double-cropping corn after wheat is riskier than double-cropping sorghum because the recommended planting date of corn is from early April to the end of May. The wheat growing season overlaps significantly with optimal corn planting dates. Conversely, sorghum fits better for double-cropping after wheat, because optimal sorghum planting dates are significantly later than for corn.

Crop sequences that fit best relative to optimal planting and harvesting dates, i.e., double-cropping sorghum after wheat or wheat after corn, rather than the reverse, hold from the perspective of crop water use as well. Theoretical daily water requirements of corn and sorghum based on studies of the water use of both crops (Assefa et al., 2010; Stone et al., 2006) are presented in Figure 7.11. Because corn is planted early, its water use is greater during June and July but is less by the end of August compared with sorghum, which allows corn to leave more water for the next wheat crop than sorghum, with greatest water use in July and August. This conclusion agrees with farmer observations obtained via the survey discussed earlier. Conversely, wheat uses more water in spring, making

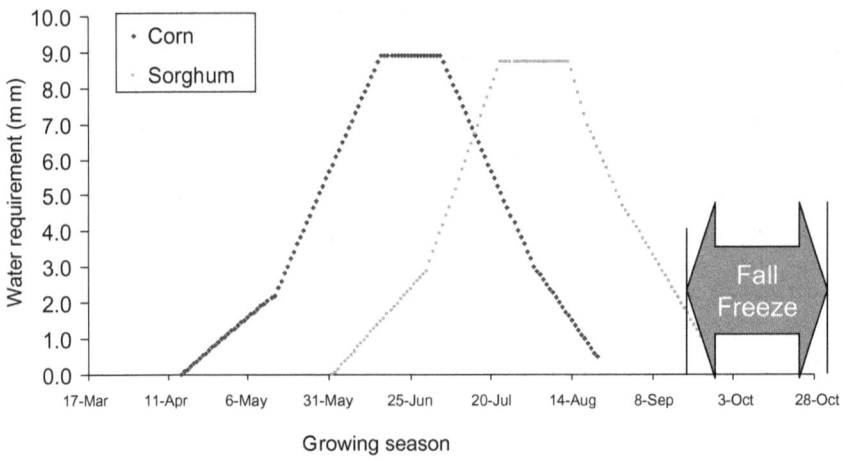

Figure 7.11 Theoretical corn (planted April 15) and sorghum (planted June 1) water use in growing season and expected range of first freeze in Kansas.

sorghum the better choice for double-crop planting after wheat harvest without irrigation.

Allelopathic effects of sorghum on following wheat (Roth et al., 2000) and on following corn (Schmidt and Frey, 1988) have been reported. Even though other independent research results indicated water-soluble extracts from sorghum residue inhibiting seed germination (Guenzi and McCalla, 1966; Guenzi et al., 1967), allelopathic effects reported from field studies, such as the report by Roth et al. (2000), are more speculative based on performance of following crops rather than direct measures of allelopathic compounds identified in field conditions. This potential for allelopathic effects is another reason why double-cropping wheat after corn is preferable to double-cropping after sorghum.

Including an extended fallow period between harvest of corn or sorghum and planting wheat (W-C-F or W-GS-F) eliminates the planting/harvesting date overlap and minimizes differences in water availability at wheat planting. About 9–11 months elapse between corn and sorghum harvest and wheat planting. Studies by Norwood (2000) and Nielson et al. (2002) showed no significant difference in soil profile water content at wheat planting in W-C-F, W-S-F, W-F, and W-Millet-F rotations. Water use should not be a basis for comparing crop rotations of wheat after corn and sorghum with an extended fallow between summer crop harvest and wheat planting.

Our main objective was to compare the fitness of corn and sorghum in Great Plains crop rotations. Based on the survey results, available moisture and pest management options were the most important factors dictating crop sequence decisions. Rotations of both corn and sorghum with wheat, soybean, and with each other were common among respondents. The most common rotation sequences among respondent farmers were W-C/GS-F, W-W-C/GS-SB, and C-GS-SB.

A summary of rotation studies revealed that wheat and soybean yields did not differ after corn compared with after sorghum in 7 out of 10 studies. The remaining three studies showed that wheat and soybean yields were greater after corn than after sorghum. No study showed greater wheat or soybean yield after sorghum compared with after corn.

Compatibility of corn and sorghum with winter wheat based on planting and harvesting dates and water use indicated that planting wheat after corn is preferred over planting after sorghum due to: (i) significant overlaps in planting and harvesting dates of wheat and sorghum, and (ii) possible differences in soil profile moisture available at wheat planting, a significant issue for farmers' choice of crop sequence. Conversely, double-cropping sorghum after wheat is preferred over double-cropping corn after wheat for similar reasons.

CHAPTER 8

General Summary

Corn and grain sorghum are among the top cereal crops worldwide, and both are key for global food security. Similarities between the two crops, particularly their adaptation for warm-season grain production, pose an opportunity for comparisons to inform appropriate cropping decisions. The main objective of this book is to provide a comprehensive comparison between corn and grain sorghum. We have reviewed available literature and analyzed datasets on corn and sorghum morphology, physiology, yield relations, resource use, and effects on cropping systems. Reviews suggested that grain sorghum and corn are morphologically similar, at least aboveground and in the vegetative growth stages. Their physiology and developmental stages also are similar in resource-rich environments. Notable differences reported for the two crops in morphology, physiology, and phenology were related to their adaptation to different levels of stress conditions. Historical data from the USDA for both corn and sorghum suggest that corn harvest area is on the rise, whereas sorghum harvest area is declining in the United States. In general, the mean and the maximum possible yields of corn were greater than that of grain sorghum. On the other hand, the variation in dryland yield was less for sorghum than for corn. We found the variation in yield explained by environment to be higher for corn than for sorghum; i.e., corn was more responsive and sensitive to environmental variation than sorghum. The environmental conditions that correlate highly with yields of corn and sorghum were found in different portions of the growing season. Land use based comparison suggested that about 6 Mg ha^{-1} is the cutoff yield value; i.e., if corn's expected yield is above 6 Mg ha^{-1}, it has better land use, but if expected yield of corn for an area or season is below this cutoff value, sorghum has better land use efficiencies. About 533 mm of ET was found to be the cutoff above which corn, and below which sorghum, has better water use efficiency. In rainfed production, approximately 432 mm of total seasonal rainfall (April through September) was the threshold value above which corn, and below which sorghum, has better average rainfall use efficiency. Fertilizer use efficiency of corn was found to be greater than for sorghum at all application rates, as long as water is not limiting. Sorghum postemergent herbicides were unavailable, more expensive, or less effective compared to postemergent herbicides options for corn. Analysis of available research data on the effects of corn and sorghum on wheat or soybean yield in equivalent 2-, 3-, and 5-year rotation systems found no difference in 7 out of 10 studies, but greater wheat or soybean yields following corn in three studies. Rotational studies indicated that more nitrogen (N) fertilizer was required for equivalent or better yield when wheat was planted after sorghum compared with planting after corn. In summary, even though corn and sorghum appear to compete for summer cropping area, they take advantage of different parts of the season, and thus should be considered alternative crops to utilize different environmental conditions. Breeding should focus on increasing maximum yield level for sorghum and decreasing corn yield sensitivity for variations in environmental conditions, to increase competence of these crops.

A general summary is presented in Table 8.1. In this table, based on our review and analysis, we have indicated apparent advantages of

Table 8.1 A General Summary of Apparent Advantages of Corn and Sorghum Over One Another, Based on Review and Analysis of Published Research

Criterion	Corn Advantage Over Sorghum	Sorghum Advantage Over Corn
Morphology	– Better aboveground biomass to intercept radiation – Yields better than sorghum when water is not limiting	– Better underground biomass to intercept water – Yields better than corn when water is highly limiting
Physiology	– More responsive to differences in plant density – Seedlings tolerate cold better than sorghum	– Maintains physiological activities at higher level than corn in lower water conditions
Phenology	– Highly predictable growth and development based on heat unit accumulation	– Has the plasticity to hasten or withhold phenological events under water stress conditions
Trend	– An increase in yield for the past 70 years, 1939 through 2009, for both dryland and irrigated corn – On average, the rate of yield increase was higher for corn (irrigated) than for irrigated sorghum	– Average dryland yield increased for grain sorghum, between 1957 and 2008, but no significant yield changes were observed in irrigated sorghum yields

(Continued)

Table 8.1 (Continued)

Criterion	Corn Advantage Over Sorghum	Sorghum Advantage Over Corn
	– Dryland corn harvest area began to increase in about the mid-1990s and has surpassed the area occupied by dryland sorghum in the east, northeast, and west-central districts of Kansas since the mid-2000s	
Yield	– Mean and maximum yields of corn are higher than grain sorghum	– The variation in dryland yield has been less for sorghum than corn
	– Corn yield responds more to different levels of management or environmental conditions than sorghum	– Sorghum yield is less sensitive to different levels of environmental variation than corn
	– Irrigated yields of corn are almost always superior to irrigated sorghum yields.	
	– June and July rainfall seems more important for determining dryland corn yield – A warmer and dryer April correlates positively with dryland corn yield – A cooler but wetter June correlates positively with corn yield – Warmer and wet August correlates positively with corn yield	– July and August rainfall seems more important for dryland sorghum yield – A cooler and wetter April correlates positively with dryland sorghum yield – A warmer and slightly wet June correlates positively with sorghum yield – A cooler and wetter August is more beneficial to dryland sorghum
Resource use efficiency (land, rainfall, nutrient, and pesticide use efficiency)	– In environments where expected yield is above 6 Mg ha^{-1}, corn has better land use efficiencies	– In environments where expected yield is below 6 Mg ha^{-1}, sorghum has better land use efficiencies
	– Corn rainfall use efficiency is better than sorghum in rainfed production where >432 mm of total seasonal rainfall (April through September) is available	– Sorghum rainfall use efficiency is better than corn in rainfed production where expected total seasonal rainfall (April through September) is <432 mm
	– Nitrogen fertilizer use efficiency of corn is better than sorghum at any equivalent nutrient supply level as long as water is not limiting	
	– Corn has better postemergent herbicide choices that are relatively inexpensive and effective	

(*Continued*)

Table 8.1 (Continued)

Criterion	Corn Advantage Over Sorghum	Sorghum Advantage Over Corn
Rotation (compatibility in cropping system)	— Effects of corn on yield of subsequent crops in common rotation systems were better than or similar to that of sorghum	
	— Planting wheat in the same season after corn harvest was more compatible than wheat after sorghum	— Double-cropping sorghum rather than corn after wheat harvest better optimized water use and available growing season

either corn or sorghum over one another according to several criteria. Because these advantages are based mainly on review of literature and analysis of data collected for purposes other than direct corn–sorghum comparisons, some overgeneralization may be possible. The two crops also have multiple hybrids, some of which might deviate from the general statements herein. These comparisons, we hope, will indicate future research direction (for both crops), help policy decisions, educate students, and open further research ideas.

REFERENCES

Abbe, E.C., Stein, O.L., 1954. The origin of the shoot apex in maize: embryogeny. Am. J. Bot. 41, 285–293.

Abendroth, L.J., Elmore, R.W., Boyer, M.J., Marlay, S.K., 2011. Corn growth and development. PMR1009, Iowa State University Extension, Ames, IA.

Ackerson, R.C., Kreig, D.R., 1977. Stomatal and nonstomatal regulation of water use in cotton, corn, and sorghum. Plant Physiol. 60 (6), 850–853.

Anderson, E.L., 1987. Corn root growth and distribution as influenced by tillage and nitrogen fertilization. Agron. J. 79, 544–549.

Anderson, R.L., Bowman, R.A., Nielsen, D.C., Vigil, M.F., Aiken, R.M., Benjamin, J.G., 1999. Alternative crop rotations for the central Great Plains. J. Prod. Agric. 12, 95–99.

Artschwager, E., 1948. Anatomy and morphology of the vegetative organs of *Sorghum vulgare*. USDA Technical Bulletin No. 957, Washington, DC.

Assefa, Y., Staggenborg, S.A., 2010. Grain sorghum yield with hybrid advancment and changes in agronomic practices from 1957 through 2008. Agron. J. 102, 703–706.

Assefa, Y., Staggenborg, S.A., 2011. Phenotypic changes in grain sorghum over the last five decades. J. Agron. Crop. Sci. 197 (4), 249–257.

Assefa, Y., Staggenborg, S.A., Prasad,V.P.V., 2010. Grain sorghum water requirement and responses to drought stress: a review. Online, Crop Management. Available from: http://dx.doi.org/10.1094/CM-2010-1109-01-RV.

Assefa, Y., Roozeboom, K.L., Staggenborg, S.A., Du, J., 2012. Dryland and irrigated corn yield with climate, management, and hybrid changes from 1939 through 2009. Agron. J. 104, 473–482.

Atteya, A.M., 2003. Alteration of water relations and yield of corn genotypes in response to drought stress. Bulg. J. Plant Physiol. 29 (1–2), 63–76.

Barber, S.A., 1972. Relation of weather to the influence of hay crops on subsequent corn yields on a Chalmers silt loam. Agron. J. 64 (1), 8–10.

Barker, T., Campos, H., Cooper, M., Dolan, D., Edmeades, G., Habben, J., et al., 2005. Improving drought tolerance in maize. Plant Breed. Rev. 25, 173–253.

Beadle, C.L., Stevensor, K.R., Neumann, H.H., Thurtell, G.W., King, K.W., 1973. Diffuse resistance, transpiration, and photosynthesis in single leaves of corn and sorghum in relation to leaf water potential. Can. J. Plant Sci. 53, 537–544.

Benson, G.O., 1985. Why the reduced yields when corn follows corn and possible management responses. In: Proceedings of the Corn and Sorghum Research, Chicago, IL, pp. 971–972.

Blum, A., 2009. Effective use of water (EUM) and not water-use efficiency (WUE) is the target of crop yield improvement under drought stress. Field Crops Res. 112, 119–123.

Blum, A., Arkine, G.F., Jordan, W.R., 1977. Sorghum root morphogenesis and growth I. effect of maturity genes. Crop Sci. 17, 149–153.

Blum, A., Ritchie, J.T., 1984. Effect of soil surface water content on sorghum root distribution in the soil. Field Crops Res. 8, 169–176.

Blumenthal, J.M., Lyon, D.J., Stroup, W.W., 2003. Optimal plant population and nitrogen fertility for dryland corn in western Nebraska. Agron. J. 95, 878–883.

Boyer, J.S., 1970. Differing sensitivity of photosynthesis to low leaf water potentials in corn and soybeans. Plant Physiol. 46, 819–820, IRRI, Philippines.

Bruulsema, T.W., Tollenaar, M., Heckman, J.R., 2000. Boosting crop yields in the next century. Better Crops 84, 9–13.

Campbell, C.S. 2012. Poaceae. Encyclopædia Britannica Online Academic Edition. Available online at: <http://www.britannica.com/EBchecked/topic/465603/Poaceae>.

Cardwell, V.B., 1982. Fifty years of Minnesota corn production: source of yield increase. Agron. J. 74, 984–990.

Castleberry, R.M., Crum, C.W., Krull, F., 1984. Genetic yield improvement of U.S. maize cultivars under varying fertility and climatic environments. Crop Sci. 24, 33–36.

CFIA.,1994. The biology of *Zea mays* (L.). Biology Document, Plant Biosafety Office, Canada Food Inspection Service.

Claassen, M.M., 2003. Effect of nitrogen rate and seeding rate on no-till winter wheat after grain sorghum. Kansas Fertilizer Research Report 921, Kansas State University, pp. 69–71.

Claassen, M.M., 2006. Reduced tillage and crop rotation system with wheat, grain sorghum, corn and soybean. Field Research Report of Progress 992, Kansas State University, pp. HC3–HC8.

Claassen, M.M., Regehr, D.L., 2009. No-till crop rotation effects on wheat, corn, grain sorghum, soybeans and sunflower. Field Research Report of Progress 1017, Kansas State University, pp. 36–43.

Clapp, A.L., Jugenheimer, R.W., Hollembeak, H.D., Skold, L.N., 1939. Kansas Corn Tests. Kansas State University Agricultural Experiment Station, Manhattan, KS.

Compton, L.L., 1943. Moisture conservation practices and the relationship of conserved water to crop yields. SSSAJ 7, 368–373.

Schlenker, W., Roberts, M., 2009. Nonlinear temperature effects indicate severe damages to U.S. crop yields under climate change. PNAS 106, 15594–15598.

Conley, S.P., Stevens, W.G., and Dunn, D.D., 2005. Grain sorghum response to row spacing, plant density, and planter skips. Online, Crop Management. Available from: http://dx.doi.org/10.1094/CM-2005-0718-01-RS.

Creswell, R., Martin, F.W., 1998. Dryland Farming: Crops and Techniques for Arid Regions. ECHO, North Fort Myers, FL.

Dick, W.A., Vandoren, J.D.M., 1985. Continuous tillage and rotation combinations effects on corn, soybean, and oat yields. Agron. J. 77 (3), 459–465.

Dillon, S.L., Shapter, F.M., Henry, R.J., Cordeiro, G., Izquierdo, L., Lee, S., 2007a. Domestication to crop improvement: genetic resources for sorghum and saccharum (Andropogoneae). Ann. Bot. 100, 975–989.

Dillon, S.L., Lawrence, P.K., Henry, R.J., Price, H.J., 2007b. Sorghum resolved as a distinct genus based on combined ITS1, ndhF and Adh1 analyses. Pl. Syst. Evol. 268, 29–43.

Duch-Carvallo, T., Malaga, J., 2009. International sorghum trade: United States beyond the Mexican dependency? Southern Agricultural Economics Association Annual Meeting, Atlanta, GA, January 31–February 3.

Duvick, D.N., 1977. Genetic rates of gain in hybrid maize yields during the past 40 years. Maydica 22, 187–196.

Duvick, D.N., 1984. Genetic contributions to yield gains of U.S. hybrid maize, 1930–1980. In: Genetic Contributions to Yield Gains of Five Major Crop Plants, CSSA Special Publication No. 7.

Duvick, D.N., 1999. Heterosis: feeding people and protecting natural resources. The Genetics and Exploitation of Heterosis in Crops. American Society of Agronomy and Crop Science Society of America, Inc., Madison, WI, pp. 19–29.

Duvick, D.N., 2005. The contribution of breeding to yield advances in maize. Adv. Agron. 86, 83–145.

Duvick, D.N., Cassman, K.G., 1999. Post-green revolution trends in yield potential of temperate maize in the north-central United States. Crop Sci. 39, 1622–1630.

Edmeades, G.O., Tollenaar, M., 1990. Genetic and cultural improvements in maize production. In: Sinha, S.K., Sane, P.V., Bhargava, S.C., Agrawal, P.K. (Eds.), Proceedings of the International Congress of Plant Physiology. Society for Plant Physiology and Biochemistry, New Delhi, India, pp. 164–180.

Eghball, B., Power, J.F., 1995. Fractal description of temporal yield variability of 10 crops in the United States. Agron. J. 87, 152–156.

FAO, 1999. The Future of Our Land: Facing the Challenge. Land and Water Development Division, FAO, Rome. Available online at: <http://www.fao.org/docrep/004/x3810e/x3810e04.htm>.

Fischer, K.S., Johnson, E.C., Edmeades, G.O., 1982. Breeding and selection for drought resistance in tropical maize. Drought Resistance in Crops with Emphasis on Rice. International Rice Research Institute, Laguna, Philippines.

Fisher, N.M., Dunham, R.J., 1984. Root morphology and nutrient uptake. In: Goldsworthy, P.R., Fisher, N.W. (Eds.), Physiology of Tropical Field Crops. John Wiley & Sons, New York, NY, pp. 85–117.

Frank, B.J., 2010. Corn grain yield and plant characteristics in two water environments. Masters Thesis, Kansas State University.

Garcia, P., Offutt, S.E., Pinar, M., Changnon, S.A., 1987. Corn yield behavior: effect of technology advance and weather conditions. J. Clim. Appl. Met. 26, 1092–1102.

Ghannoum, O., von Caemmerer, S., Ziska, L.H., Conroy, J.P., 2000. The growth response of C_4 plants to rising atmospheric CO_2 partial pressure: a reassessment. Plant Cell Environ. 23, 931–942.

Gordon, W.B., 1991. Tillage and Fall–Spring nitrogen combination for wheat production following early-season corn. Field Research Report of Progress 655, Kansas State University, pp. 50–53.

Gordon, W.B., Staggenborg, S.A., 2003. Comparing corn and grain sorghum in diverse environments. Field Research Report of Progress 913, Kansas State University Agricultural Experiment Station, Manhattan, KS.

Guenzi, W.D., McCalla, T.M., 1966. Phenolic acids in oats, wheat, sorghum, and corn residues and their phytotoxicity. Agron. J. 58, 303–304.

Guenzi, W.D., McCalla, T.M., Norstadt, F.A., 1967. Presence and persistence of phytotoxic substances in wheat, oat, corn, and sorghum residues. Agron. J. 59, 163–165.

Hallauer, A.R., 1973. Hybrid development and population improvement in maize by reciprocal full-sib selection. Egypt. J. Genet. Cytol. 2, 84–101.

Halvorson, A.D., Nielsen, D.C., Reule, C.A., 2004. Nitrogen fertilization and rotation effects on no-till dryland wheat production. Agron. J. 96, 1196–1201.

Hamman, L., Boland, M., Dhuyvetter, K.C., 2002. Economic issues with grain sorghum. Agricultural Market Resource Center, Kansas State University. Available online at: <http://www.agmrc.org/media/cms/economicissuessorghum_0CED92AD4D3FE.pdf>.

Hannaway, D.B., Myers, D., 2004. Forage fact sheet: sorghum. Available online at: <http://forages.oregonstate.edu/php/fact_sheet_print_grass.php?SpecID=24&use=Forage>.

Hansen, J., Sato, M., Ruedy, R., Lo, K., Lea, D.W., Medina-Elizade, M., 2006. Global temperature change. Proc. Nat. Acad. Sci. USA 103, 14288–14293.

Hansen, N., Allen, B., Baumhardt, R.L., Lyon, D., 2012. Research achievements and adoption of no-till, dryland cropping in the semi-arid U.S. Great Plains. Field Crops Res. 132, 196–203.

Heer, W.F., 1999. Effect of nitrogen rate on yield in continuous wheat and wheat in alternative crop rotation in south central Kansas. Field Research Report of Progress 854, Kansas State University, pp. 134–138.

Heer, W.F., 2009. Effect of nitrogen rate and previous crop on grain yield in continuous wheat and alternative cropping system in south central Kansas. Field Research Report of Progress 1031, Kansas State University, pp. 53–64.

Hesketh, J., 1967. Enhancement of photosynthetic CO_2 assimilation in the absence of oxygen, as dependent upon species and temperature. Planta 76, 371–374.

Hesketh, J.D., Chase, S.S., Nanda, D.K., 1969. Environmental and genetic modification of leaf number in maize, sorghum, and Hungarian millet. Crop Sci. 9, 460–463.

Hochholdinger, F., Woll, K., Sauer, M., Dembinsky, D., 2004. Genetic dissection of root formation in maize (Zea mays) reveals root-type developmental programmes. Ann. Bot. 93, 359–368.

Hochholdinger, F., Woll, K., Sauer, M., Feix, G., 2005. Functional genomic tools in support of the genetic analysis of root development in maize (Zea mays L.). Maydica 50, 437–442.

Hodgest, T., Kanemasu, E.T., Teare, I.D., 1979. Modeling dry matter accumulation and yield of grain sorghum. Can. J. Plant Sci. 59, 803–818.

Hoppe, D.C., McCully, M.E., Wenzel, C.L., 1986. The nodal roots of Zea: their development in relation to structural features of the stem. Can. J. Bot. 64, 2524–2537.

Hornbeck, R., 2012. The enduring impact of the American dust bowl: short- and long-run adjustments to environmental catastrophe. Am. Econ. Rev. 102 (4), 1477–1507.

Howell, T.A., Steiner, J.l., Schneider A.D., Evett, S.A. and Tolk, J.A., 1994. Evaporation of irrigated winter wheat, sorghum and corn. ASAE international summer meeting, June 19–22, 19944. Kansas city, KS.

Houghton, J.T., Jenkins, G.J., Ephraums, J.J. (Eds.), 1990. Climate Change: The IPCC Scientific Assessment. Cambridge University Press, Cambridge (Intergovernmental Panel on Climate Change, World Meteorological Organization, United Nations Environmental Program).

House, L.R., 1985. A Guide to Sorghum Breeding. International Crops Research Institute for the Semi-Arid Tropics, ICRISAT, Patancheru, Andhra Pradesh, India, p. 13.

Howell, T.A., 2001. Enhancing water use efficiency in irrigated agriculture. Agron. J. 93, 281–289.

Hu, Q., Buyanovsky, G., 2003. Climate effects on corn yield in Missouri. J. Appl. Meteor. 42, 1626–1635.

Kansas State University, 1939–2009. Kansas Corn Performance Trial Reports. Kansas State University Agricultural Experiment Station and Cooperative Extension Service Bulletin and Report of Progress, Manhattan, KS.

Kansas State University, 1957–2008. Kansas Grain Sorghum Performance Trial Reports. Kansas State University Agricultural Experiment Station and Cooperative Extension Service Bulletin and Report of Progress, Manhattan, KS.

Kaplan, D.R., 2001. The science of plant morphology: definition, history, and role in modern biology. Am. J. Bot. 88 (10), 1711–1741.

Kaylen, M.S., Koroma, S.S., 1991. Trend, weather variables, and the distribution of U.S. corn yields. Rev. Agric. Econ. 13 (2), 249–258.

Kelley, J., 2006. Sorghum growth and development. In: Espinoza, L. and Kelley, J. (Eds.), Grain Sorghum Hand Book. Cooperative Extension Service, University of Arkansas. Available at: <http://www.uaex.edu/Other_Areas/publications/PDF/MP297/MP297.pdf>.

Kelley, K.W., Sweeney, D.W., 2009. Effect of previous crop, nitrogen placement method, and time of nitrogen application on no-till wheat yield. Field Research Report of Progress 1031, Kansas State University, pp. 76–77.

Kershner, K.S., Al-Khatib, K., Krothapalli, K., Tuinstra, M.R., 2012. Genetic resistance to acetyl-coenzyme a carboxylase-inhibiting herbicides in grain sorghum. Crop Sci. 52, 64–73.

Kiesselbach, T.A., 1999. The Structure and Reproduction of Corn. Cold Spring Harbor Laboratory Press, Cold Spring, NY.

Kim, S.H., Sicher, R.C., Baew, H., Gitzz, D.C., Baker, J., Timlin, D.J., et al., 2006. Canopy photosynthesis, evapotranspiration, leaf nitrogen, and transcription profiles of maize in response to CO_2 enrichment. Global Change Biol. 12, 588–600.

Klocke, N., Currie, R., 2009. Corn and grain sorghum production with limited irrigation. In: Proceedings of the 21st Annual Central Plains Irrigation Conference, Colby, KS, February 24–25.

Klocke, N.L., Currie, R.S., Dumler, T.J., 2009. Water saving from crop residue management. In: Proceedings of the 21st Annual Central Plains Irrigation Conference, Colby, KS, pp. 74–80.

Klocke, N.L., Currie, R.S., Tomsicek, D.J., Koehn, J., 2011. Corn yield response to deficit irrigation. Trans. ASABE 54 (3), 931–940.

Knowles, T.C., Hipp, B.W., Graff, P.S., Marshall, D.S., 1993. Nitrogen nutrition of rainfed winter wheat in tilled and no-till sorghum and wheat residues. Agron. J. 85, 886–893.

Koch, E., Bruns, E., Chmielewski, F.M., Defila, C., Lipa, W., Menzel, A., 2007. Guidelines for plant phenological observations. Available online at: <http://www.cluster.bom.gov.au>.

Kramer, P.J., Boyer, J.S., (Eds.), 1995. Stomata and gas exchange. In: Kramer, P.J., Boyer, J.S. Water Relations of Plants and Soils. Academic Press, London, pp. 257–282.

Kucharik, J.C., Serbin, S.P., 2008. Impacts of recent climate change on Wisconsin corn and soybean yield trends. Environ. Res. Lett. 3, 1–10.

Kuleshov, M.N., 1933. World's diversity of phenotypes of maize. Agron. J. 25, 688–700.

Lambin, E.F., Meyfroidt, P., 2011. Global land use change, economic globalization, and the looming land scarcity. PNAS 108 (9), 3465–3472.

Lamm, F.R., Aiken, R.M., Kheira, A.A., 2009. Corn yield and water use characteristics as affected by tillage, plant density, and irrigation. Trans. ASABE 52 (1), 133–143.

Larson. K., Thompson, D., Harn, D., 2001. Limited and full irrigation comparison for corn and grain sorghum. Available online at: <http://www.colostate.edu/depts/prc/pubs/LimitedandFullIrrigationComparisonforCorn.pdf>.

Leakey, A.D.B., 2009. Rising atmospheric carbon dioxide concentration and the future of C_4 crops for food and fuel. Proc. R. Soc. B 276, 2333–2343.

Leff, B., Ramankutty, N., Foley, J.A., 2004. Geographic distribution of major crops across the world. Global Biogeochem. Cycles 18, GB1009. Available from: http://dx.doi.org/10.1029/2003GB002108.

Leikam, D.F., Lamond, R.E., Mengel, D.B., 2003. Soil test interpretation and fertilizer recommendations. Kansas State University Agricultural Experiment Station and Cooperative Extension Service, MF-2586.

Lemaire, G., Hébert, C.Y., 1996. Nitrogen uptake capacities of maize and sorghum crops in different nitrogen and water supply conditions. Agronomie 16, 231–246.

Li, X., Takahashi, T., Suzuki, N., Kaiser, H.M., 2011. The impact of climate change on maize yields in the United States and China. Agric. Syst. 104 (4), 348–353.

Lingenfelser, J., Adee, E., Aschwege, W., Bond, D., Evans, P., Heer, W., et al., 2012. Kansas performance tests with corn hybrids. Agricultural Experiment Station Report of Progress 1073, Kansas State University, Manhattan, KS.

Lobell, D.B., Asner, P.G., 2003. Climate and management contributions to Recent Trends in U.S. agricultural yields. Science 299 (5609), 1032.

Maghsoudi, K., Moud, A.M., 2008. Analysis of the effects of stomatal frequency and size on transpiration and yield of wheat (*Triticum aestivum* L.). Am.-Euras. J. Agric. Environ. Sci. 3 (6), 865–872.

Martin, J.H., 1930. The comparative drought resistance of sorghum and corn. Agron. J. 22, 993–1003.

Martin, J.N., 1920. Botany with Agricultural Applications. John Wiley & Sons, New York, NY.

Mason, S.C., Delon, K., Eskridge, K.M., Galusha, T.D., 2008. Yield increase has been more rapid for maize than for grain sorghum. Crop Sci. 48, 1560–1568.

Mathews, O.R., Brown, L.A., 1938. Winter wheat and sorghum production in the southern Great Plains under limited rainfall. USDA Circular 477.

Mayaki, W.C., Stone, L.R., Teare, I.D., 1976. Irrigated and nonirrigated soybean, corn, and grain sorghum root systems. Agron. J. 68, 532–534.

McConnell, R.I., Gardner, C.O., 1979. Selection for cold germination in two corn populations. Crop Sci. 19, 765–768.

Menz, K.M., Pardey, P., 1983. Technology and U.S. corn yields: plateaus and price responsiveness. AAEA 65 (3), 558–562.

Metz, B., Davidson, O.R., Bosch, P.R., Dave, R., Meyer, L.A., 2007. Climate Change 2007: Mitigation. Cambridge University Press, Cambridge, New York, NY (Contribution of Working Group III to the Fourth Assessment Report of the Intergovernmental Panel on Climate Change).

Miller, E.C., 1916. Comparative study of the root systems and leaf areas of corn and sorghum. J. Agric. Res. 6, 311–332.

Mock, J.J., Bakri, A.A., 1976. Recurrent selection for cold tolerance in maize. Crop Sci. 16, 230–233.

Mock, J.J., Eberhart, S.A., 1972. Cold tolerance in adapted maize populations. Crop Sci. 12, 466–469.

Moroke, T.S., Schwartz, R.C., Brown, K.W., Juo, A.S.R., 2005. Soil water depletion and root distribution of tree dryland crops. Soil Sci. Soc. Am. J. 69, 197–205.

Morrison, J., Morikawa, M., Murphy, M., Schulte, P., 2009. Water scarcity and climate change: global risk for business and investors. Pacific Institute, Oakland, CA. Available online at: <http://www.ceres.org/resources/reports/water-scarcity-climate-change-risks-for-investors-2009>.

Muchow, R.C., Sinclair, T.R., 1989. Epidermal conductance, stomatal density and stomatal size among genotypes of *Sorghum bicolour* (L.) Moench. Plant Cell Environ. 12, 425–431.

Musick, J.T., Dusek, D.A., 1980. Irrigated corn yield response to water. Am. Soc. Agric. Eng. Trans. 23, 92–98.

Nakayama, F.S., van Bavel, C.H.M., 1963. Root activity distribution patterns of sorghum and soil moisture conditions. Agron. J. 55, 271–274.

NASS, 2007. USDA National Agricultural Statistics Service. USDA-NASS, Washington, DC.

Neild, R.E., Newman, J.E., 1987. Growing season characteristics and requirements in the Corn Belt. National Corn Handbook, NCH-40.

Nelson, W.L., Dale, R.F., 1978. Effect of trend or technology variable and record period on production of corn yield with weather variables. J. Appl. Metro. 17, 926–933.

Nielsen, D.C., Vigil, M.F., Anderson, R.L., Bowman, R.A., Benjamin, J.G., Halvorson, A.D., 2002. Cropping system influence on planting water content and yield of winter wheat. Agron. J. 94, 962–967.

Nishimoto, R.K., Warren, G.F., 1971. Shoot zone uptake and translocation of soil applied herbicides. Weed Sci. 19 (2), 156–161.

Norwood, C.A., 1999. Water use and yield of dryland row crops as affected by tillage. Agron. J. 91, 108–115.

Norwood, C.A., 2000. Dryland winter wheat as affected by previous crops. Agron. J. 92, 121–127.

Norwood, C.A., 2001. Dryland corn in western Kansas: effects of hybrid maturity, planting date, and plant population. Agron. J. 93, 540–547.

Ojima, S.D., Lackett, J.M., 2002. Preparing for a changing climate: the potential consequences of climate variability and change. Report for the US Global Change Research Program, Colorado State University, p. 103.

Osuna-Ortega, J., Mendoza-Castillo, M., del, C., Medoza-Onofre, L.E., 2003. Sorghum cold tolerance, pollen production, and seed yield in the central High Valleys of Mexico. Maydica 48, 125–132.

Ottman, M.J., Kimball, B.A., Pinter, P.J., Wall, G.W., Vanderlip, R.L., Leavitt, S.W., et al., 2001. Elevated CO_2 increases sorghum biomass under drought conditions. New Phytol. 150, 261–273.

Oury, B., 1965. Allowing for weather in crop production model building. J. Farm Econ. 47 (2), 270–283.

Peterson, T.A., Varvel, G.E., 1989. Crop yield as affected by rotation and nitrogen rate. I. Soybean. Agron. J. 81, 727–731.

Pimentel, D., Hepperly, P., Hanson, J., Douds, D., Seidel, R., 2005. Environmental, energetic, and economic comparisons of organic and conventional farming systems. BioScience 55 (7), 573–582.

Poorter, H., Roument, C., Campbell, B.D., 1996. Interspecific variation in the growth response of plants to elevated CO_2: a search for functional types. In: Körner, C., Bazzaz, F.A. (Eds.), Carbon Dioxide, Populations, and Communities. Academic Press, San Diego, CA, pp. 375–411.

Quinby, J.R., Hesketh, J.D., Voigt, R.L., 1973. Influence of temperature and photoperiod on floral initiation land leaf number in sorghum. Crop Sci. 13, 243–246.

Rees., J., Irmak, S., 2012. Crop water use comparison of rainfed corn, sorghum, and soybean from 2009 to 2011. Nebraska Crop Production and Pest Management Information (article ID 4835538). Available online at: <http://cropwatch.unl.edu/web/cropwatch/archive?articleID=4835538>.

Ritchie, S.W., Hanway, J.J., Benson, G.O., 1996. How a Corn Plant Develops. Iowa State University of Science and Technology Cooperative Extension Service, Ames, IA.

Roozeboom, K.L., Rife, C.R., Vanderlip, R.L., 1994. Corn and grain sorghum trial stand establishment methods. Agronomy Abstracts, ASA, Madison, WI.

Rosegrant, M.W., Cai,X., Cline, S.A., 2002. Global water outlook to 2025: averting an impeding crises. International Food Research Institute, Washington, DC, and International Water Management Institute, Colombo, Sri Lanka. Available online at: <http://www.ifpri.cgiar.org/sites/default/files/pubs/pubs/fpr/fprwater2025.pdf>.

Roth, C.M., Shoyer, J.P., Paulsen, G.M., 2000. Allelopathy of sorghum on wheat under several tillage systems. Agron. J. 92, 855–860.

Runge, E.C.A., Odell, R.T., 1958. The relation between precipitation, temperature and the yield of corn on the Agronomy South Farm, Urbana, Illinois. Agron. J. 50, 448–454.

Russell, W.A., 1974. Comparative performance of maize hybrids representing different eras of maize breeding. In: Proceedings of the 29th Annual Corn and Sorghum Research Conference, Chicago, IL, December 10–12. American Seed Trade Association, Washington, DC, pp. 81–101.

Sadras, V.O., Grassini, P., Steduto, P., 2007. Status of water use efficiency of main crops. SOLAW Background Thematic Report-TR07, FAO.

Sanchez-Diaz, M.F., Karmer, P.J., 1971. Behavior of corn and sorghum under water stress and during recovery. Plant Physiol. 48, 613–616.

Sanderson, F.H., 1954. Methods of Crop Forecasting. Harvard University Press, Cambridge, MA, p. 259.

Scarsbrook, C.E., Doss, B.D., 1973. Leaf area index and radiation as related to corn yield. Agron. J. 65, 459–461.

Schlegel, A., Dumler, T., Holman, J., Stone, L., 2010. No till and crop rotation in western Kansas. Available online at: <http://www.wkarc.org/doc11065.ashx>.

Schlegel, A.J., 2006a. Long-Term Nitrogen and Phosphorus Fertilization on Yield of Irrigated Corn. Kansas State University, Manhattan, KS. Available online at: <http://www.wkarc.org/doc3165.ashx>.

Schlegel, A.J., 2006b. Long-Term Nitrogen and Phosphorus Fertilization on Yield of Irrigated Grain Sorghum. Kansas State University, Manhattan, KS. Available online at: <http://www.wkarc.org/doc3170.ashx>.

Schlegel, A.J., Frickel, D.L., 1994. Nitrogen management in a wheat–sorghum–fallow rotation. Field Research Report of Progress 719, Kansas State University, pp. 36–39.

Schlegel, A.J., Dhuyvetter, K.C., Schaffer, J.A., 1994. Effect of a previous soybean crop and nitrogen fertilizer on irrigated corn and grain sorghum. Field Research Report of Progress 719, Kansas State University, pp. 22–27.

Schlenker, W., Roberts, M., 2009. Nonlinear temperature effects indicate severe damages to U.S. crop yields under climate change. PNAS, 106, pp. 15594–15598.

Schlenker, W., Hanemann, M., Anthony, F., 2004. The impact of global warming on U.S. agriculture: an econometric analysis of optimal growing conditions. Department of Agricultural and Resource Economics, Berkeley, CA.

Schmidt, G., Frey, E., 1988. Crop rotation effects in northern Ghana. In: Unger, P.W., Sneed, T.V., Jordan, W.R., Jensen, R. (Eds.), Proceedings of the Challenges in Dryland Agriculture Conference, August 15–19, Bushland, TX. Texas Agricultural Experiment Station, College Station, TX, pp. 775–777.

Seiler, R.A., 1984. A yield model for grain sorghum on the basis of crop responses to prevailing weather and climatic conditions. Diss. Abstr. Int. Sci. Eng. 44 (12), 3604.

Shackel, K.A., Hall, A.E., 1984. Effect of intercropping on the water relations of sorghum and cowpea. Field Crops Resea. 8, 381–387.

Shaw, L.H., 1964. The effect of weather on agricultural output: a look at methodology. J. Farm Econ. 46 (1), 218–230.

Shroyer, J.P., Ohlenbusch, P.D., Duncan, S., Thompson, C., Fjell, D.L., Kilgore, G.L., et al., 1996. Kansas crop planting guide. Kansas State University Agricultural Experiment Station and Cooperative Extension Service, L-818.

Sieglinger, J.B., 1920. Temporary roots of the sorghums. Agron. J. 12, 143–145.

Sieglinger, J.B., 1936. Leaf number of sorghum stalks. Agron. J. 28, 636–642.

Sinclair, T.R., Tanner, C.B., Bennett, J.M., 1984. Water-use efficiency in crop production. BioScience 34 (1), 36–40.

Singh, M., Ogren, W.L., Widholm, J.M., 1974. Photosynthetic characteristics of several C_3 and C_4 plant species grown under different light intensities. Crop Sci. 14, 566–568.

Singh, V., Van Oosterom, E.J., Jordan, Messina, C.D., Cooper, M., Graeme, L., et al., 2010. Morphological and architectural development of root systems in sorghum and maize. Plant Soil 333, 287–299.

Slife, F.W., 1976. Economics of herbicide use and cultivar tolerance to herbicide. In: Proceedings of the Annual Corn Sorghum Research Conference, 7–9 December, 1976, Chicago, IL. American Seed Trade Association, Washington, DC, pp. 77–82.

Smith, C.W., Frederiksen, A., 2000. History of cultivar development in the United States: from "Memories of A. B. Maunder—Plant Breeder". Sorghum: Origin, History, Technology, and Production. John Wiley & Sons, New York, NY, pp. 191–225.

Smith, J.W., 1903. Relationship of precipitation to yield of corn. USDA Yearbook. USDA, Washington, DC, pp. 215–224.

Staggenborg, S.A., Whitney, D.A., Fjell, D.L., Shroyer, J.P., 2003. Seeding and nitrogen rates required to optimize winter wheat yields following grain sorghum and soybean. Agron. J. 95, 253–259.

Staggenborg, S.A., Dhuyvetter, K.C., Gordon, W.B., 2008. Grain sorghum and corn comparisons: yield, economic, and environmental responses. Agron. J. 100, 1600–1604.

Stanger, T.F., Lauer, J.G., 2006. Optimum plant population of Bt and non-Bt corn in Wisconsin. Agron. J. 98, 914–921.

Stone, L.R., Schlegel, A.J., 2006. Yield–water supply relationships of grain sorghum and winter wheat. Agron. J. 98, 1359–1366.

Stone, L.R., Schlegel, A.J., Khan, A.H., Klocke, N.L., Aiken, R.M., 2006. Water supply: Yield relationships developed for study of water managment. J. Nat. Resour. Life Sci. Educ. 35, 161–173.

Stone, L.R., Schlegel, A.J., Lamm, F.R., Spurgeon, W.E., 1994. Storage efficiency of preplant irrigation. J. Soil Water Conserv. 49 (1), 72–76.

Stone, L.R., Schlegel, A.J., Gwin, R.E., Khan, A.H., 1996. Response of corn, grain sorghum, and sunflower to irrigation in the High Plains of Kansas. Agric. Water Manage. 30 (3), 251–259.

Stone, L.R., Goodrum, D.E., Schlegel, A.J., Jaafar, M.N., Khan, A.H., 2002. Water depletion depth of grain sorghum and sunflower in the central High Plains. Agron. J. 94, 936–943.

Sung, F.J., Krieg, D.R., 1979. Relative sensitivity of photosynthetic assimilation and translocation of carbon to water stress. Plant Physiol. 64, 852–856.

Tannura, M.A., Irwin, S.H., Good, D.L., 2008. Weather, technology, and corn and soybean yields in the U.S. Corn Belt. Marketing and Outlook Research Report, Department of Agricultural and Consumer Economics, University of Illinois at Urbana-Champaign.

Tarkalson, D.D., Hergert, G.W., Cassman, K.G., 2006a. Long-term effects of tillage on soil chemical properties and grain yields of a dryland winter wheat–sorghum/corn–fallow rotation in the Great Plains. Agron. J. 98, 26–33.

Tarkalson, D.D., Payero, J.O., Hergert, G.W., Cassman, K.G., 2006b. Acidification of soil in a dry land winter wheat–sorghum/corn–fallow rotation in the semiarid U.S. Great Plains. Plant Soil 283, 367–379.

Teare, I.D., Peterson, C.J., Law, A.G., 1971. Size and frequency of leaf stomata in cultivars of *Triticum aestivum* and other *Triticum* species. Crop Sci. 11, 496–498.

Thompson, C.R., Peterson, D.E., Fick, W.H., Stahlman, P.W., Wolf, R.E., 2009. Chemical weed control for field crops, pastures, rangeland, and noncropland. Field Research Report of Progress 1081, pp. 24–45, Kansas State University, Manhattan, KS.

Thompson, L.M., 1969. Weather and technology in the production of corn in the U.S. Corn Belt. Agron. J. 61, 453–456.

Thompson, L.M., 1975. Weather variability, climatic change, and grain production. Science 188 (4188), 535–541.

Tokatlidis, I.S., 2013. Adapting maize crop to climate change. Agron. Sustain. Dev. 33, 63–79.

Tolk, J.A., Howell., T.A., 2008. Field water supply: yield relationships of grain sorghum grown in three USA southern Great Plains soils. AGWAT-2627.

Troyer, A.F., 2000. Temperate corn: background, behavior, and breeding. In: Hallauer, A.R. (Ed.), Specialty Corns. CRC press, Washington, DC, pp. 393–466.

Turner, N.C., 1974. Stomatal behavior and water status of maize, sorghum and tobacco under field conditions. Plant Physiol. 53, 360–365.

Unger, P.W., Baumhardt, R.L., 1999. Factors related to dryland grain sorghum yield increases, 1939 through 1997. Agron. J. 91, 870–875.

US Environmental Protection Agency (EPA), 1998. Climate change and Kansas. EPA 236-F-98-007i.

USDA, 1930–1990. Agricultural Statistics Service. U.S. Government Printing Office, Washington, DC.

USDA-NASS, 2009. Crop Production Historical Track Records. U.S. Government Printing Office, Washington, DC.

USDA, 2011. Crop Production Historical Track Records. U.S. Government Printing Office, Washington, DC.

USGS (US Geological Survey), 2001. Selected findings and current perspectives on urban and agricultural water quality by the National Water-Quality Assessment program. US Department of the Interior, Washington, DC.

Vanderlip, R.L., 1993. How a Sorghum Plant Develops. Kansas State University Agricultural Experiment Station and Cooperative Extension Service, Manhattan, KS.

Wade, L.J., Douglas, A.C.L., 1990. Effect of plant density on grain yield and yield stability of sorghum hybrids differing in maturity. Aust. J. Exp. Agric. 30 (2), 257–264.

Warrington, I.J., Kanemasu, E.T., 1983. Corn growth response to temperature and photoperiod. III. Leaf number. Agron. J. 75, 762–766.

Wary, R.E., Whitney, D.A., Lamond, R.E., Kilgore, G.L., 1994. Effect of nitrogen rates on wheat following grain sorghum, wheat, and soybeans. Kansas Fertilizer Research Report 719, Kansas State University, pp. 12–14.

Watt, M., Magee, L.J., McCully, M.E., 2007. Types, structure and potential for axial water flow in the deepest roots of field grown cereals. New Phytol. 178, 135–146.

Weatherwax, P., 1916. Morphology of the flower of Zea Mays. Bull. Torrey Bot. Club 43, 127–144.

Weaver, J.E., 1926. Root Development of Field Crops. McGraw-Hill, New York, NY.

Wenzel, C.L., McCully, M.E., Canny, M.J., 1989. Development of water conducting capacity in the root systems of young plants of corn and some other C_4. Plant Physiol. 89, 1094–1101.

Wet, J.M.J.D., Harlan, J.R., 1971. The origin and domestication of *Sorghum bicolor*. Econ. Bot. 25 (2), 128–135.

Whiteman, P.C., Wilson, G.L., 1965. Effects of water stress on the reproductive development of *Sorghum vulgare* Pers, 4. University of Queensland Papers, Queensland, Australia, pp. 233–239.

Wicks, G.A., Stahlman, P.W., Anderson, R.L., 1995. Weed management systems for semiarid areas of the central Great Plains. Proc. North Central Weed Sci. Soc. 50, 74–190.

Wilhelm, E.P., Mullen, R.E., Keeling, P.L., Singletary, G.W., 1999. Heat stress during grain filling in maize: effect on kernel growth and metabolism. Crop Sci. 39, 1733–1741.

Yu, J., Tuinstra, M.R., Claassen, M.M., Gordon, W.B., Witt, M.D., 2004. Analysis of cold tolerance in sorghum under controlled environment conditions. Field Crops Rese. 85, 21–30.

Zelitch, I., 1971. Photosynthesis, Photorespiration, and Plant Productivity. Academic Press, New York, NY.

Zhang, W., 2000. Phylogeny of the grass family (Poaceae) from *rpl16* intron sequence data. Mol. Phylogenet. Evol. 15 (1), 135–146.

www.ingramcontent.com/pod-product-compliance
Lightning Source LLC
Chambersburg PA
CBHW071406290426
44108CB00014B/1713